All business leaders should read leadership quality but it has been, for most of us, a serious missing link in our development. This book is not only a great gift but also highlights our great responsibility because—as it clearly demonstrates—change is possible.
Tex Gunning, CEO, TNT Express

Sander Tideman and the Dalai Lama give clear and useful insights into triple value creation in business. They powerfully remind us that leading a business is about creating social, economical, and ecological value for every individual, every organization, and for our society. Their insights combined with Sander's personal journey will stimulate your intrinsic human motivation to implement these valuable and practical insights immediately.
Lorike Hagdorn, Professor in Business Administration, Vrije Universiteit Amsterdam

Sander has captured the deeply human approach of the Dalai Lama in relation to business and its role in society. Through wonderful dialogues and inspiring visions, the book generates a spirit of hope by providing tangible, practical insights. And it is not just for Buddhists—this is a book for everyone who cares for people and the planet!
Peter Blom, CEO and Chairman of the Executive Board, Triodos Bank Group

I rejoice that Sander Tideman has been able to capture our dialogues with the Dalai Lama in this book. This will greatly help us to understand that, in order for us to develop humanity in the secular world, we need to integrate compassion into our economy and business.
Ruud Lubbers, Former Prime Minister, the Netherlands

Sander Tideman's book is a soaring testimony from a lifelong seeker, wisdom teacher, and business leader. He provides insight across a breathtaking range, from the underpinnings of Adam Smith's "invisible hand" to the redemptive power of the human heart. Sander draws us a map for the positive future of business at a time when we desperately need inspiration and guidance on where to go next. I recommend this brilliant work to anyone who takes their role seriously, be it in business, in society, or—as Sander so passionately teaches us—at the inescapable nexus of them both.
Erica Ariel Fox, author of *New York Times* bestseller, *Winning From Within: A Breakthrough Method for Leading, Living, and Lasting Change*

Sander Tideman's lifelong commitment to building a more compassionate and socially responsible "values-based" business world shines through in this compendium of dialogues with His Holiness the Dalai Lama. This book is a must-read for those who possess a humanistic world vision, strong societal empathy, and for those who seek to make the 21st century one that has "social responsibility" at the center of the global business ecosystem.
Peter Miscovich, Managing Director, Jones Lang LaSalle, New York

This book is a testimony to Sander's many gifts. He has managed to weave his diverse and broad insights into a powerful intellectual framework and practical guide for making shared purpose the new modus operandi of business. This is, I believe, the path for both personal and business success. ... I recommend this book to anyone who is seeking to understand how business can help create real solutions to address the enormous global challenges and opportunities we face.
Anders Ferguson, Founding Principal, Veris Wealth Partners (from the foreword)

Finally—a deep, thoughtful, and constructive book that connects our characters, values, and intentions with our leadership and entrepreneurship practices. Too often we fool ourselves into believing that we can achieve a just, inclusive, and sustainable economy without first practicing justice, inclusion, and respect towards all living things. Sander Tideman's lifelong search, which he generously shares in these pages, is uniquely positioned to be an invaluable guide for all those serious in this quest.
Marcello Palazzi, Founder, Progressio Foundation and B Lab Europe

As someone who has spent more than a decade and a half watching the work of Sander Tideman and his colleagues unfold, I am thrilled with this inspiring, original, and far-sighted business book. ... The major achievement of this work, in my view, is that it is the best business explanation of the connection between inner and outer transformation I've ever seen.
David L. Cooperrider, PhD, University Distinguished Professor, Case Western Reserve University (from the afterword)

This impressive book reminds us of the essentials of life and business. Without purpose there is no clear identification for leaders, employees, and customers to be successful. Sander Tideman's own journey, full depth, and richness of experience and content, illustrate very well how we are all challenged every day. The book gives a lot of inspiration of how we all can find a more purposeful journey.
Thomas Burbl, CEO and Chairman, AXA Group

Business as an Instrument for Societal Change: In Conversation with the Dalai Lama

In Conversation with
the Dalai Lama

Business as an Instrument for
Societal Change

Sander G. Tideman

Greenleaf
PUBLISHING

© 2016 Sander Tideman and the Flow Impact Fund

Published by Greenleaf Publishing Limited
Salts Mill, Victoria Road, Saltaire, BD18 3LA, UK
www.greenleaf-publishing.com

Cover photo: Dalai Lama by Christopher Michel

Printed in the UK on environmentally friendly, acid-free paper
from managed forests by CPI Group (UK) Ltd, Croydon.

British Library Cataloguing in Publication Data:
    A catalogue record for this book is available from the British Library.

    ISBN-13: 978-1-78353-450-0 [hardback]
    ISBN-13: 978-1-78353-452-4 [paperback]
    ISBN-13: 978-1-78353-451-7 [PDF ebook]
    ISBN-13: 978-1-78353-517-0 [ePub ebook]

Dedicated to my grandfather Sander van Deinse,
who taught me an apprecation for business,
and my father Paul Tideman, for inspiring
me with an interest in serving society

Contents

Part 1:
Compassion or competition

Part 2:
Designing an economy that works for everyone

Part 3:
Leadership for a sustainable world

Part 4:
Education of the heart

Figures and tables

Figures

Tables

Foreword

His Holiness, the Dalai Lama

THE DALAI LAMA

Many people tell me that the world is facing a moral crisis. Standards of living have improved greatly over the last hundred years. However, global economic development has also led to a widening of the gap between rich and poor and contributed to climate change. This, in turn, is leading to widespread damage to the natural environment, threats to food security, and increasing scarcity of potable water; problems that will affect us all. Unattended, these difficulties will continue to worsen, but we cannot leave it to governments and politicians to find solutions. Business leaders have a special responsibility because they have knowledge and resources to address these issues.

I know that business and making money are important, but I have no idea how to do either. However, I do know that for any human endeavor to be constructive, good motivation is required. We need a new approach to solving problems, which implies a new approach to ethics. To promote change, it is important that education promotes fundamental human values, which, coupled with human intelligence, can be the basis for creating a better world.

The focus of business should not be to cheat, exploit, or deceive others, nor even just to generate profit. Business should also contribute to society at large. To think that cultivating concern for others means neglecting our own interests is a mistake. I believe it is in the interest of business to adopt a broader, more far-sighted perspective that takes account of everyone involved. There needs to be a greater respect for the oneness of human beings, the sense that we all belong to one human family. Viewing others as human beings who are like us is a kind of humility. It does not mean we cannot be firm when necessary; it is not a question of being meek. For example, when a company is more transparent it attracts greater trust and respect, which makes it more successful in the long term.

Over recent years, I have engaged in dialogue with business leaders and economists to discuss these questions. It is exhilarating to see how many of them are exploring fresh and alternative approaches to their work. My friend Sander Tideman, who initiated many of these dialogues and has tried to adopt a more ethical and principled approach in his own work as a banker and business consultant, has gathered these conversations together in this book.

If the focus of business is to change from merely maximizing profits to maximizing social well-being and sustainable development that protects the environment, then business leaders will need to broaden their outlook and learn new skills. I hope that the ideas and discussions recounted in these pages will serve to inspire that change.

September 8, 2015

Foreword

Anders Ferguson

Imagine a world in which business is not only a driver of economic growth, but also a builder of sustainable societies and the steward of the environment. A place where mindful business leaders understand that profit is just one part of their responsibility. A place in which executives truly believe that long-term prosperity is possible only when communities and ecosystems are stable, healthy, and equitable.

For those of us who dream of a better way, welcome to the world of Sander Tideman. My dear friend and colleague for nearly two decades, Sander is a visionary thinker who has written a book that may be among the most important you will read, both personally and professionally.

Sander's work is particularly special to me because of our long partnership, exploring and connecting sustainability with business leadership, spiritual wisdom, and science. How can they be integrated to create truly inspired enterprises? Sander has taken our shared work, including our dialogues with the Dalai Lama, and woven it with his keen mind and expansive research into a powerful model for creating sustainable value through developing what he calls "shared purpose." It is my hope this book will inspire you to change yourself—and perhaps a bit of the world too.

I have spent my life in business, primarily focused on sustainability. Over the years, I have come to realize that the positive changes in business that I worked to create depended on deeper inner knowledge and inner leadership. I began exploring mindfulness and meditation. I turned to the Dalai Lama's teachings for insight.

In 2000, I went on retreat to a holy mountain on the old border of Tibet and China. Separated from all that was familiar and with a quieter mind, I had a moment of insight: for the sake of the planet, executives need to develop inner leadership that is focused as much on creating common good as on our own immediate business interests. Those charged with running businesses, I thought, can have a positive and holistic impact on society if guided by mindfulness, whole-systems approaches, and the radical idea of altruistic motivation.

I returned home with new hope and perspective. Inspired by the Dalai Lama, I thought perhaps he could help bring together leaders to explore real change. I learned that Sander Tideman, in the Netherlands, had organized just such a meeting in 1999 in Amsterdam. He called it "Compassion or Competition."

Sander and I first met in New York on a bright Saturday afternoon in November 2000. When we started talking, it was as if I had found an old friend. It was a pivotal moment in my life.

Sander was trained as a lawyer and banker. He had long practiced meditation, which informed his work. He had met the Dalai Lama many years earlier, and it very much seemed as though we were seeking answers to the same questions. How do business leaders respond to the challenges of sustainability? How do leaders prepare themselves and cultivate their inner leadership to address these increasingly complex problems? How do we see and create breakthrough innovations? How can we evolve a more humane capitalism, one in which business advanced the greater good?

My answer to those questions came in the form of a seemingly crazy idea which I posed to Sander shortly after we met: we should host a world conference of business leaders, spiritual leaders, scientists, and thought leaders with the Dalai Lama. Sander loved the idea. To his great credit, he signed up to the project and the magic began.

Sander invited his friend and collaborator, Marcello Palazzi, from Europe and I assembled change agents from the U.S.A. Together, this amazing team convened a global conference called "Spirit in Business: Ethics, Mindfulness and the Bottom Line" in New York City in the spring of 2002. It was remarkably well received and was one of those breakthrough events that truly created new ways of knowing and seeing. Our success, in no small part, was due to the 600 leaders who joined us from 35 countries. Amazingly, the conference was sponsored by forward-thinking global firms such as American Express, Forbes, and Verizon.

Over the next five years, Sander and I were blessed to work with truly remarkable leaders from across the globe, producing conferences and research-learning networks. We met with inspired companies, researchers, and leaders everywhere. We were lucky enough to be part of co-creating an emerging body of knowledge about new sustainable approaches to managing companies. I have little doubt that more than a decade later the thousands of us who have been touched by Spirit in Business are making positive changes, whether in Global 100 companies or local family-owned businesses.

We also agreed to convene "Compassion or Competition" dialogues with the Dalai Lama every few years to continue the exploration. The learnings and wisdom of these dialogues are captured in this book. In fact, it is the grounded and profound insights of the Dalai Lama that Sander used as the foundation for developing his theory of change, which is at the heart of this brilliant book.

Also captured in the book are insights from those early days of our thinking. Sander expertly integrates these ideas in the coming pages:

- Neuroscience, psychology, quantum physics, and branches of science are translating ancient wisdom traditions into new ways of knowing. Science is increasingly the language of business.

- Interconnectedness and interdependence is not just a cool idea, but one of the underpinnings of successful global companies. Whole-systems approaches connect all stakeholders.

- Leaders searching for deeper transformative practices, such as compassion and mindfulness, behave differently and can take more responsibility. As the Dalai Lama has pointed out, the most important point of meditation is to develop a positive, altruistic motivation, as a source of action. Balanced minds make better decisions with better effects.

Working together with Sander has been transformative. Our early years also prepared both of us for the next phase of our lives—parallel paths that would bring Spirit in Business to more people.

Sander's journey deepened and expanded the practice of high-impact business leadership and research. In the process, he developed a model for how business is becoming an agent of positive social change. I focused on high-impact investing in sustainable companies by cofounding Veris Wealth Partners with a group of industry pioneers. We created one of the world's leading firms devoted to sustainable and impact investing. This discipline has become one of the most effective ways to direct capital to the greatest challenges facing our planet.

So what else can we all learn from this book? In truth, there are many profound lessons and observations, but the most exciting for me is Sander's emerging paradigm for a flourishing business.

Sander makes crystal clear the essential interconnectedness of business, the environment, and humanity—ideas that the Dalai Lama speaks about passionately on an almost daily basis. Sander's central thesis is that leaders who follow their enlightened self-interest can grow great and profitable businesses while being catalysts for social change. In doing so, they can strengthen communities and preserve the beauty of the nature.

Sander constructs a robust framework for how business can accomplish this objective by moving beyond its seemingly single focus on the profit motive. The key, he writes, is creating shared value on three levels: value for the organization itself; value for customers; and value for society as a whole. He refers to this as "triple value".

A world in which triple value is the central organizing principle for business is a revolutionary idea, and yet it is not as far off as it may seem. Global business leaders are becoming aware that they can no longer operate outside the context of their societies and without regard for the environment. As Feike Sijbesma, CEO of DSM, is quoted in this book: "How can you be successful as a business in a society that fails?"

Sander posits that the process of creating shared purpose sustainably demands new mindsets. Put simply, if we as leaders find our own altruistic purpose, we have the authenticity to shape the deepest purpose for our company. Embedding this shared purpose for the common good becomes a deliberate business strategy in which executives, shareholders, and customers are increasingly aligned and often well rewarded.

Over the past 15 years, research increasingly indicates that companies implementing sustainable business practices outperform those that don't. The stock performance of publicly traded companies using sustainable models and environmental, social, and governmental (ESG) practices is growing stronger than those that don't. Some of the most notable outperformers are those in which women play meaningful leadership roles at the board and the executive level. Interestingly, this was an emerging trend we identified in the early days of Spirit in Business research. These companies are truly building shared purpose.

No one should kid themselves that creating shared purpose will be easily implemented. However, Sander emphasizes its possibility if interconnectedness is at the core of the new business mind. If business leaders are willing to take the risk of transforming their worldview, accepting new mindsets and concepts, we can break from the past. We will only succeed in creating a new world if we look through new eyes, eyes that better connect our minds with our hearts and with the whole.

Drawing on neuroscience and research, Sander makes the compelling case that every business leader is also part of a species in which care, altruism, and responsibility are interwoven into our DNA. Genuine cooperation was standard operating procedure until societies were atomized and splintered by division of labor. Sander's message is: we need to get our mojo back and think of ourselves as one tribe again. Shared purpose refocuses us on our common good.

This book is a testimony to Sander's many gifts. He has managed to weave his diverse and broad insights with his professional business experience, creating a powerful intellectual framework and practical guide for making shared purpose the new modus operandi of business. This is, I believe, the path for both personal and business success.

It is rare in life to find such a committed colleague and kindred spirit. I am extremely grateful to Sander for playing such a grounding and visionary role in encouraging me on this path. I regard this book as an outcome of our shared journey and recommend it to anyone who is seeking to understand how business can help create real solutions to address the enormous global challenges and opportunities we face.

Anders Ferguson, Founding Principal, Veris Wealth Partners
New York City, July 2016

Acknowledgments

There are many people who I would like to thank for their contribution to this book. First and foremost is H.H. the Dalai Lama, for generously giving time to the dialogues and being a constant source of inspiration to me. I am also grateful to the dozens of other leaders who lend their wisdom to the dialogues. Their names are mentioned at the end of this book.

The dialogues in the Netherlands have been made possible by the Dalai Lama Netherlands Foundation, especially my fellow board members Tsering Jampa, Paula de Wijs-Koolkin, Florens van Canstein, Marco Werre, and Reinier Tilanus. We received special support from Ruud Lubbers, former prime minister of the Netherlands, and Erica Terpstra, a Dutch parliamentarian. The first public dialogue in Amsterdam was co-created by Marcello Palazzi of the Progression Foundation, with whom I went on to cofound Spirit in Business, together with Anders Ferguson, in order to bring these dialogues into the world.

Anders Ferguson then took the lead in organizing the second dialogue with the Dalai Lama at the University of California Irvine, co-hosted by Spirit in Business. We received support from University California Irvine's Vice Chancellor, Manuel Gomez, and the philanthropists Lori and Bob Warmington. Anders regularly reminded me of the significance of maintaining these dialogues, which he continued to support. He also kindly offered a foreword to this book as a token of our shared endeavor.

The third dialogue in The Hague was co-hosted by the Global Leaders Academy, especially my partners Sue Cheshire and Jeroen Drontmann, and Focus Conferences, led by Hans Groen. Important contributions came from Princess Irene van Lippe Biesterfeld, Ruud Lubbers, and Herman Wijffels.

The fourth dialogue with the Dalai Lama in Rotterdam was co-created with Rob van Tulder, Eva Rood, and Joep Elemans of Rotterdam School of Management, Erasmus University, who also hosted the event at their campus. We received support from Garrison Institute and Flow Foundation. Daniel Siegel, Ger Rombouts, and Muriel Arts deserve special mentions.

Deirdre Taylor and Louise Tideman helped to transcribe the proceedings of the dialogues. Rebecca Marsh of Greenleaf Publishing, who quickly saw the potential of turning the proceedings into a book, provided constant encouragement, insisting that I share my personal story in this book. Others who encouraged me in my attempt to integrate Buddhist wisdom with modern management science are B. Alan Wallace, Reinout van Lennep, Joe Loizzo, Diana Rose, and Daniel Siegel. My research partner Muriel Arts contributed significant ideas and models to Chapters 9 and 10 of the book. Flow Impact Fund, which Muriel Arts and I co-founded, provided the necessary funds so that I could take time off to write the book. Lambert van der Aalsvoort and Sarah Wong offered their photographs, while Hemels van der Hart (Henk, Pieter, and Yorick) created the cover and illustrations. David Cooperrider completed the book by writing an inspirational afterword. Finally I thank my family who traveled with me on this journey: my dear Sandra and beautiful children Louise, Laurens, and Bas.

Introductory quotations from the Dalai Lama

The new reality is one in which the challenges facing humanity are "beyond individual effort" and our interdependences have become even starker. The gaps between our perception and the new reality are based on having concepts from the previous century in our minds, and this wrong perception creates the wrong approach. In business, these 20th-century concepts include management approaches that focus on maximizing short-term profits, ignoring activists or critical stakeholders, only caring about a regulatory license and an "us and them" approach to problem-solving.

Companies are living, complex organisms and not profit machines. The profit should therefore not be the object of a company, but rather a result of good work. Just like a person can't survive for long without food and water, a company can't survive without profits. But just as we cannot reduce the purpose of a human to eating and drinking alone, we cannot regard companies solely as money-making entities.

We live in one world, but our mind sees differences.

Introduction

There are a number of books that explore the need to incorporate ethical and sustainable practice in business and an ever-growing number of books that help individuals achieve greater fulfillment in their lives. This book sets out to be different: it interweaves the story of a lawyer/banker who was hardwired into commonplace notions of economic and personal success (me!) and the life choices that I have made over the course of my career, with the wider story of what is happening in the global economy and business world. This story is punctuated by a number of meetings with one of the world's most respected leaders, H.H. the Dalai Lama. Through dialogue between the Dalai Lama and a number of notable thinkers in business and society, important questions are raised, each reflecting a period in time both in my own career and in broader global events.

The discussion, however, also transcends any specific points in time and provides some important questions for humanity: can personal motivation and fulfillment be aligned with economic and business prosperity? What does economic and business prosperity mean in the long term? Should we change our notion of prosperity to not only ensure the long-term viability of the planet and social equality, but also improve the sense of happiness and well-being in individuals? Can the purpose of individuals, business, and society be aligned?

The book contains the transcripts of four public meetings with the Dalai Lama, starting in 1999, on a topic that the Buddhist leader does not often speak on: the role of business and economics in society. These meetings took place over two decades in which the world has seen massive change, both positive and negative. The reach and impact of business economically, environmentally, and socially has become immense—employing 3.1 billion people, with 400 million earning less than US$1.25/day, generating 60% of global GDP, and employing 70% of all workers. Globally, 125 million small and medium-size businesses exist, and multinational companies have revenues similar in size to GDPs of entire nations. The amount of wealth and prosperity that has been created in the last few decades is staggering.

At the same time, largely due to the very role of economic activity, the world is facing multiple crises, ranging from growing economic and social inequality, to climate change, biodiversity loss, and natural resource depletion. The economic crises that erupted in 2008 caused the general public to lose trust in business, which remains dishearteningly low. Recent surveys report that only one in four members of the general public trusts business leaders to rectify issues, and only one in five trusts them to tell the truth and make ethical and moral decisions. The 2014 Edelman Trust Barometer, a 27-country survey with more than 33,000 respondents, found that overall trust has declined across countries and sectors; in terms of credible spokespeople to win public trust, CEOs ranked second-lowest at 43% and government officials the lowest at 36%.[1]

Perhaps the obstacle to trust remains the practice of business; not least, the view that the role of business and managers is to optimize the return to one stakeholder only—namely, the shareholder. Yet discussions on capitalism revert to oft-repeated arguments on the role of corporations as necessary agents of wealth and employment, a discussion perpetuated by business schools and regulators, assuming business and society are disconnected.

Throughout the discussions, the Dalai Lama, in his usually modest way, does not profess to understand the world of business fully nor does he force a single point of view on his audience. However, his insightful comments act as a guide. They speak of the need to cultivate compassion, long-term thinking, and creativity, and to change the rules of the game. The critics of this book may dismiss the learning as unworkable in a world that is based on competition, market forces, and winning at all cost. However, as the dialogues contained in this book demonstrate, this type of thinking will not help us at a time when the world is facing some of the most challenging political, economic, and ecological questions that impact on whole societies and the planet.

1   Edelman Berland (2014). 2014 Edelman Trust Barometer. Retrieved from http://www.edelman.com/insights/intellectual-property/2014-edelman -trust-barometer.

While the goals of economics and business are generally understood as generating the material conditions for a comfortable life, it is increasingly clear that these goals cannot be achieved if we continue to pursue economic growth as the overarching goal of human life. The problem is that the dominant economic model overlooks the fact that our planet simply has insufficient natural resources for unlimited growth, just as it fails to recognize the need for developing key internal conditions such as care and reducing mental and emotional suffering. Therefore, it is both necessary and urgent that we broaden our outlook on the purpose of business and economics.

Inspiration for new perspectives

After each public dialogue with the Dalai Lama, the participants went home with inspiration and new perspectives, precisely because they enable us to pause and reflect. Looking back at what was discussed, these exchanges are not only a tribute to the natural wisdom that springs from the Dalai Lama, but they do in fact carry relevance for the major challenges that confront our leadership in business and economics today. All dialogues close with the conclusion that these challenges—ranging from climate change to population growth, poverty, natural resource depletion, unemployment, lack of trust, and financial sector transformation—cannot be overcome by conventional responses from business and governments, such as more/less market regulation, more/less monetary incentives or austerity, or more/less consumer spending. Rather, these challenges require a leadership response that comes from tapping more deeply into our qualities as human beings: our courage, our dreams, our sense of hope, or, as the Dalai Lama simply put it, our heart. All these are features of the human mind—the very domain in which the Dalai Lama is considered an expert.

These dialogues can be traced back to a chance encounter I had with the Dalai Lama when I was a young student traveling through India in 1982. This book recounts the initial discussions I had with the Tibetan Buddhist leader, because—looking back over more than 30 years—the questions we discussed have bearings on many questions faced by humanity today. With hindsight, I can say that these questions became the starting point for a lifelong inquiry into the relationship between economics and happiness, between business and well-being, and between money and values. My subsequent career in international law, banking, and consulting—working with many leaders in business and society—did not give me many more answers. On the contrary, it caused me to be ever more puzzled by the discrepancies between the promise and practice of economic theory.

What I saw was that, even though tremendous financial wealth was created in the business world, little of this seemed to trickle down to those at the bottom of the pile, especially in developing countries. And everywhere I traveled, East and West, North and South, I observed that economic development went hand in hand with environmental degradation. So, while economics focused on building financial capital, it simultaneously had an erosive effect on social and natural capital. The financial crisis in 2008 and subsequent economic recession finally convinced me—and many others— that there is something fundamentally wrong in the way we manage our economies and societies, and that we should look for a new paradigm to create value in the future.

Different paradigm for business and economics

While the dialogues with the Dalai Lama may initially generate more questions than answers to the many choices that face us, they can be regarded— as they are regarded throughout this book—as steps toward the development of a new economic and business paradigm. This paradigm will uproot the notion of the *Homo economicus*—a fundamental pillar under classical economics with its belief in free-market competition and selfish consumers. The new paradigm allows business organizations to contribute positively to society by recognizing that business is *a part of* society and not *apart from* society.

The origin of the separation between business and society is often traced back to the 18th-century philosopher Adam Smith, who said in the *Wealth of Nations*:

> It is not from the benevolence of the butcher, the brewer, or the baker that we expect our dinner, but from their regard of their won interest. We address not to their humanity but to their self-love, and never talk to them of our own necessities, but of their advantages. [2]

2   Smith, A. (2008). *The Wealth of Nations*. New York, NY: Oxford University Press. (Original work published 1776).

Subsequent economists have interpreted this as the most fundamental principle of economics: "people are actuated only of self-interest."[3] In business, this principle manifested in the idea promoted by Milton Friedman that "the only business of business is business."[4]

Two centuries later, we can see that this principle is belief or ideology only—not a fact of science. Even though the whole economics and management profession became hooked on these beliefs, they are not "real," in the sense that they are not substantiated by facts of how people behave in reality.

As numerous clinical and organizational studies have indicated, positive emotions such as compassion, care, love, and trust create profound biochemical and neurological changes in our bodies that are inherently good for us physiologically. Studies of contemplatives have revealed that we have the ability to literally transform neural pathways of our brains through mental training in care and altruism. Lower blood pressure, increased immunity, enhanced cognitive function, and improved communication skills are all examples of the types of benefit that these practices bring to the individuals and organizations practicing them. In other words, there is a definite relationship between transcending narrow self-interest through altruistic intentions, and enhanced individual and organizational performance. We all know how customers feel when a product, service, or salesperson is genuinely caring, and solves a problem with grace and care. Care makes business sense.

This being so, can we transcend current economic orthodoxy and create a more caring and holistic economy? Can we acknowledge the interdependence of business and society, that one cannot flourish without the other? Can we start to see that positive human feelings and qualities make remarkably good business sense? Can businesses help create a society in which their customers and stakeholders would wish to live?

The key to understanding how the societal performance of business can be improved lies in recognizing the important role of human motivation. Just as every human being excels when he is driven by a sense of purpose in his work, organizations too need to be *purposeful*, with *purpose* defining the "why" of business activity. An intentional and broadened focus on

3   See for example: Edgeworth, F.Y. (1967). *Mathematical Psychics: An Essay on the Application of Mathematics to the Moral Science*. New York, NY: Augustus M. Kelley Publishers. (Original work published 1881).

4   Friedman, M. (1970, September 13). The social responsibility of business is to increase its profits. *New York Times*, p. SM17.

purpose—the reason for which business is created or exists, its meaning and direction—can help organizations to develop a larger capacity for contributing to solving societal challenges.

The dialogues recounted in this book call for greater attention to the purpose of organizations and their leaders. Specifically, they call for leaders of business to develop a type of purposeful leadership so that their business will truly create societal, economic, and ecological value. This book will define that as the "societal leadership of business." If we want to develop this type of leadership, we will first need to take a step back in time and thoroughly reflect on why we are doing what we do, and whether we are moving in the right direction. As the Chinese sage Lao Tzu once said: "If you do not change direction, you may end up where you are heading."

The shared purpose between business and society

The first public meeting with the Dalai Lama was in Amsterdam in 1999 with the theme "Compassion or Competition"; the second, "Designing an Economy that Works for Everyone," in Irvine, California, in 2004; the third, "Leadership for a Sustainable World," in The Hague in 2009; and the fourth, on "Education of the Heart," in Rotterdam in 2014.

The four public dialogues—each five years apart—have followed a progression from general to more specific insights. The first dialogue in Amsterdam addressed the general role of business in society, at a time when this was hardly considered a relevant discussion. At the time of the second dialogue, the themes of corporate social responsibility (CSR) and sustainability had become more mainstream business, hence the dialogue in California focused on various aspects of making economic systems more human- and nature-friendly, or more inclusive. The third dialogue in The Hague occurred shortly after the beginning of the financial crisis, when it was evident that urgent action was needed for reconnecting business and society. It thus focused on leadership—specifically the leadership that is required for building a sustainable world. It had become clear that the necessary change could only take place when we improve our education practices. The fourth dialogue in Rotterdam, therefore, discussed the educational steps involved in cultivating the full human potential for taking leadership for a sustainable world, which we called "Education of the Heart."

The dialogues were typically designed as an opportunity to discuss a broad range of topics, without strict thematic boundaries and facilitation, so as to optimize the flow of inspiration. Nonetheless, looking back at the four dialogues, there was a recurrence of six broad themes and topics:

1. **Holistic thinking.** Modern reality is increasingly interconnected and interdependent at various levels of complexity, including the personal, the organizational, and the societal/environmental dimension. This requires a holistic approach to solving problems.

2. **Compassion.** People are naturally wired for compassion, and people in business are no exception. In fact, business cannot exist without providing services to society. As the needs in society change and grow, the role of business is to serve these needs in collaboration with many stakeholders.

3. **Responsibility.** Everybody has the capacity to develop a positive and responsible attitude, regardless of the circumstances. This attitude leads to self-confidence and courage, as well as resilience and happiness in spite of outer challenges.

4. **Creativity.** With changing reality, it becomes important to actively seek the creation of solutions. This requires leaders to develop a vision and set long-term goals, while allowing for experimentation.

5. **Collectiveness.** Since everything is increasingly interconnected, solutions for the future are beyond individual effort. This may include changing the rules of the game, new ways of measuring performance, establishing collaborative structures, and improving systems such as education.

6. **Education of the mind/heart.** It is important and possible to continuously train the mind, for example through meditation and cultivating positive qualities. The learning process involves a full engagement of our physical, emotional, mental, and spiritual dimensions, rather than mere exchange of information. In this way, levels of consciousness can be developed and grown.

At a personal level, the dialogues have helped me to gain a sense of direction in my life, which can be translated as my personal path of purpose. In the following chapters the dialogues with the Dalai Lama will be intersected with the story of my work, broken down into four phases that coincide with each dialogue. I encountered specific challenges at each stage, and experienced breakthroughs, often associated with a major shift in my career, such as leaving the world of corporate banking. I discovered that there are a number of steps involved when you try to live a life of purpose, each requiring a deepening of your commitment to personal growth.

These first stages go from "finding your purpose" (described in Chapter 1) to "applying your purpose" (described in Chapter 3). This is followed by a phase in which I could apply my purpose to the context I operated in, where my sense of purpose gained an external contextual dimension, which is described as "finding shared purpose" (described in Chapter 5). The final stage is "living shared purpose" (described in Chapter 7). When I speak of "purpose," I mean the interplay of a number of factors, including your motivation, your vision, the degree of your connectedness and competence, and—perhaps most importantly—your understanding of your operating context. This will be explained in the concluding Chapters 9 and 10.

The closing chapters of the book integrate the lessons learned from the decades of dialogue into a new framework for business performance, creating value for one's customers, for one's organization, and for society—which I call "triple value." A number of leading companies have embarked on this path, with tentative yet encouraging results. However, for this framework to be implemented, we need a new type of leadership with the ability to transform business and economics into instruments for positive societal change. This book describes this as "societal leadership in business," with six succinct leadership mind-sets and qualities. With this model, business leaders can actually create triple value with their organizations. A defining feature of this leadership is the ability to create a "shared purpose" between business and society, recognizing the various levels of interdependence between these domains that the modern economic reality presents. Chapter 10 describes the "shared purpose model," which business leaders can use to assess their own capacity for this type of leadership.

As I experienced in my own life, the development of this type of leadership corresponds to the stages in personal growth, at each stage deepening your sense of shared purpose. I believe that only with this kind of leadership can we create a more sustainable and compassionate society on this planet.

Together, the chapters paint a surprising picture of the latest possibilities to transform the way we think about business, markets, economics, and our human potential for responsibility, compassion, care, and happiness—and, most significantly—how business can become a catalyst for positive change.

Sander Tideman

Part 1:
Compassion or competition

The author, his wife Sandra, and daughter Louise, visiting the Dalai Lama in 1991.

1
The Buddha and the banker

The connection between banking and the thinking of the Dalai Lama may seem at odds, and a career in finance combined with an interest in Buddhism unusual. This introductory chapter, therefore, describes my first meeting with the Dalai Lama and the impact this has had on my outlook and career. Ultimately, it awakened a sense of "purpose" in me at a time when my own and the wider world was entirely seduced by the benefits of economic growth.

After meeting the Dalai Lama, I chose to specialize my law studies in the emerging economies of Asia and then to become a banker in China, just when it had embraced drastic free-market reforms. As a bank manager in China, I not only handled exciting business opportunities in a booming market, I also explored the economic conditions in the disadvantaged regions of Tibet and Mongolia. These explorations led to further discussion with the Dalai Lama, ten years after my first meeting. After witnessing firsthand the Asian financial crisis in 1998 and the rise of excessive focus on short-term shareholder value in the financial industry, I became gradually disillusioned with mainstream banking and the larger system of capitalism, again turning to the Dalai Lama for guidance. The discussion that ensued became the first in a series of public dialogues with him, the theme of which was "Compassion or Competition."

1.1 Meeting the Dalai Lama

In the summer of 1982, just before graduating from law school in the Netherlands and starting a career, I decided to prolong my worry-free student life with a study tour in India. Together with my student friend, Florens van Canstein, I ended up spending a few months in the Himalayas. As students of international law, we took a special interest in the Tibetan government-in-exile in Dharamsala, established by Tibetan refugees after China annexed their country in 1959. We had obtained an introduction to the Tibetans from Dr. Michael van Walt who was conducting a PhD at our university on the international legal status of Tibet. We met the leaders of the refugee community in Dharamsala in the Himalayan foothills, who shared with us grueling stories of the escape from their occupied homeland and their difficult life in exile. In one of these meetings, we befriended a Tibetan who arranged for an unexpected audience with His Holiness the Dalai Lama. He was much less known in the world before getting the Nobel Peace Prize in 1989, so receiving an audience was not as exceptional as it is now.

Nonetheless, I felt rather unprepared for this privilege, and felt nervous as we waited for what seemed a very long time in the waiting room of His Holiness' house. It was a rather unpresumptuous one-floor building overlooking the valley, encompassed by a fragrant flower garden. However, when we entered his room, my nervousness miraculously disappeared. The Buddhist leader, who laughed while pointing to my half-cut Bermuda pants, quickly put me at ease: "What happened to your pants?" he quipped. I had no words to describe the custom of leaving the rough edges hanging after cutting off half of the pants, so I started to laugh too.

When we sat down, his gentle smile and intense gaze quickly captured me. His answers to questions that would now look terribly naïve and uninformed have guided me throughout my life.

My first question centered on the issues between China and Tibet: "It seems to me that, ideologically speaking, there does not need to be conflict between China and Tibet. Marxism (at that time still China's leading ideology) and Buddhism do not seem to be fundamentally opposed to each other. There must be something in common between these two systems. Why is there disagreement?"

The Dalai Lama answered at length. "I find certain aspects of Marxism most praiseworthy from an ethical point of view, principally in its treatment of material equality and the defense of the poor against the exploitation by a minority. I believe one might say that the economic system closest to Buddhism would be a socialist economic system. Marxism is based on noble ideas such as the defense of the rights of those who are disadvantaged. But the energy given to the application of these principles is rooted in a violent hatred for the ruling classes, and that hatred is channeled into class struggle and the destruction of the ruling class. Once the ruling class is eliminated, there is nothing left to offer the people and everyone is reduced to a state of poverty. Why is this so? Because there is a total absence of compassion for certain groups of people in Marxism. So that is the big difference with Buddhism, which promotes compassion and care for all people, both rich and poor."

My next question concerned Eastern spirituality. I had seen many people in India, especially Tibetan refugees, who practiced spirituality and seemed so happy and content. They were smiling and singing in spite of the very harsh circumstances they faced in the temporary refugee camps. By contrast, I was leading a life of relative affluence but did not consider myself happy. How could that be?

His Holiness quickly unmasked my romantic notions about the East as misplaced disappointment with my own culture: "There is nothing that your culture lacks. Whether Eastern or Western, people are trying to be happy, seeking peace, living in harmony with the world. Maybe Western countries at this time have more economic wealth than in India or Tibet, but that does not make them different from people who have much less. Look at the recent demonstrations against a nuclear war in your country, in Europe. This is driven by the same wish for happiness and peace as we Tibetans. Basic human nature is the same; we are kind, peace-loving beings."

Then the conversation turned to the role of anger in Buddhism. I thought about what had happened to the Tibetans and whether they would ever remove the Chinese from their land without violent effort. Was violence totally rejected? What about a parent trying to discipline unruly children?

The Dalai Lama replied: "Anger is a destructive emotion, with negative consequences for both the subject and object. But there are circumstances possible where forceful action is allowed, to make a point clear. But the important factor here is your motivation. A father being firm to his child is generally motivated by his compassion for the child. The same applies to the Tibetans: we can be firm and steadfast in the face of Chinese unjust policies out of compassion for them. If the motivation is just anger and frustration, you'd better be very careful about your behavior. The main problem with anger is that it destroys your peace of mind. You will be much more happy when you manage to transform these negative emotions into patience, love, and compassion. Compassion leads to happiness, for yourself and others."

This exchange made a deep impression on me. It caused me to reflect on the choices we had made in the West that had enabled us to create material wealth, while at the same time forgetting the need to create social wealth—happiness. What could we learn from the East so that we would create a more balanced economy? And what could the East learn from the West, so that they could create more economic wealth? Perhaps most moving was the fact that His Holiness took me, a young and unimportant student in hippy pants, so seriously. He had given me his undivided attention, as if he saw more in me than I held possible for myself. It was the highlight of our journey through Asia that motivated me to start taking my own life and my own culture a bit more seriously.

It was the beginning of a quest for purpose: how can I make a difference in my life? As a student I had felt rather overwhelmed by the challenges in the world, but the unexpected meeting with the Dalai Lama had given me a glimpse of the possibilities in my life. The Zen Master D.T. Suzuki wrote: "Life as such has no meaning but you can give meaning to life." I embarked on a journey to give my life meaning.

1.2 Encountering economic theory

Back at university after my return from India, when I enrolled in a course in economics, I could find little that corresponded to the remarks of the Dalai Lama. The economics textbooks talked of economic laws which assumed that man naturally competes for scarce and limited natural resources. As the founder of modern economics, Adam Smith said in *The Wealth of Nations*:

> It is not from the benevolence of the butcher, the brewer, or the baker
> that we expect our dinner, but from their regard of their won interest.
> We address not to their humanity but to their self-love, and never talk
> to them of our own necessities, but of their advantages.[1]

While this statement became the root of the powerful concept of the free market, it has been misconstrued by subsequent economists to imply that human beings are selfish by nature.

---

[1] Smith, A. (2008). *The Wealth of Nations*. New York, NY: Oxford University Press. (Original work published 1776).

Classical economists tell you that it makes no sense to exert time, effort, and expense on maintaining values, if money can be made by ignoring them. In 1930, one of the great economists of the last century, Lord Keynes, wrote the following:

> we must pretend to ourselves and to everyone else that fair is foul and foul is fair; for foul is useful and fair is not. Avarice and usury and pre- caution must be our gods for a little longer still. For only they can lead us out of the tunnel of economic necessity into daylight.[2]

In Keynesian thought, which exerted a powerful influence on economists for much of the last century, ethical considerations are not merely irrelevant; they are actually a hindrance.

In one of my first classes I was told that the theory of economics was built on the image of the *Homo economicus*, defined as someone who is rational, individualistic, and driven by a desire for maximum utility, i.e., the optimal satisfaction of his needs. Since I hardly knew of anyone who would totally fit this description, I raised my arm to question the logic of the assumption. The answer, which I later heard many economists repeat, was revealing: "In economics we need the premise of the *Homo economicus* as a theoretical construct in order to make the overall economic logic work. And since this works so well, I suggest you just accept it so that we can move on."

I realized much later that it was these very assumptions that caused economics to be called a "dismal science," more based on theory than reality.[3] Economic theory, which considered itself a science of human economic behavior, had left human psychology and other social sciences outside its spectrum. It was based on the belief that people behaved rationally and that happiness was maximized by consumption and monetary wealth—an assumption that is contradicted by the social sciences. After my trip to Asia, I found it particularly difficult to reconcile this assumption with the happy faces I had seen in desolate Tibetan refugee camps.

2   Keynes, J.M. (1930). Economic possibilities for our grandchildren. In J.M. Keynes (Ed.), *The Collected Writings of John Maynard Keynes, Volume IX* (pp. 329-331). London: Macmillan.

3   Levy, D.M., & Peart, S.J. (2001). The secret history of the dismal science. Part I. Economics, religion and race in the 19th century. Library of Economics and Liberty. Retrieved from http://www.econlib.org/library/Columns/ LevyPeartdismal.html.

1.3 Finding purpose in economic development

In spite of my misgivings regarding the theory of classical economics, I was drawn to its application. In India—with its overwhelming levels of poverty—and the Tibetan refugee community, I had witnessed what results when there is no functioning economy or enterprise sector. Moreover, my journey through Asia had made me realize that countries initially thought of in the West as third-world countries—especially Taiwan, South Korea, Hong Kong, and Singapore—had quite unexpectedly generated considerable wealth through opening up their economies for international trade and investment. They were no longer regarded as third-world nations but rather "newly industrialized countries," or more exotically "Asian Tigers," on their way to becoming members of the first world. If the third world can become the first world just by reforming economic policy, I should try to understand what this entails. What was the secret to creating wealth? Was there an economic model possible that would create both financial and social wealth? That could create both money and happiness?

These questions led me to focus my studies on international economic law with a focus on Asia. This inspired me: for the first time in my life I had a sense of purpose. I had no idea what it would lead to, but I felt I was on the path to making a difference. Following my studies in Holland, I spent a year at the University of London at the School of Oriental and African Studies (SOAS) and London School of Economics (LSE), studying the legal and economic systems of the newly industrialized countries of Asia, especially China. In the mid-1980s this giant country—considered a member of the second world along with all socialist countries—was recovering from the harsh and introverted Mao era, and started to experiment with the market economy. It seemed as if we were finally growing into one "first" world.

The more I understood of the changes happening in China, the keener I became to actually work and live there. After completing my studies in London, I left for Taiwan to study the Chinese language. After one year of full immersion in the language and culture, and fascinated by the dynamics of this "Asian Tiger" economy, I took up a job as an intern at Baker & McKenzie, an international law firm with offices all over the world, including Taipei, where the Chinese lawyers welcomed this "long nose" (as they called foreigners) to help them with their growing number of international clients. Even though my role was small, as the only European in the office it was a great opportunity to observe the expanding international trade flows moving to Asia. Before long, I was called into meetings with Western firms to advise them on the law of the Republic of China, the formal name of Taiwan, which they had retained since "losing the mainland" to the communists in 1949.

1.4 Becoming a banker in China

After about two years as an aspirant lawyer working between continents, in 1989 a Taiwan-based manager of ABN AMRO Bank—the largest bank in the Netherlands—called and told me that the bank was looking for someone to manage and expand business in China. The candidate should be a Chinese-speaking Dutchman with a good understanding of business and economics. He said, "You may not know it, but you are the ideal candidate." I had never aspired to a career in banking and, since I was training to be a business lawyer, I could hardly see myself as a bank manager. But it did not take long for me to realize that this presented a tremendous opportunity to develop myself professionally and personally in the most exciting region at that time, as China's mainland was opening its huge market to the world. At the same time, it would enable me to explore more deeply the question lurking in the back of my mind since meeting the Dalai Lama: is it possible to create an economy that creates both economic and social wealth?

My banking career had begun. In the summer of 1990, I was appointed the bank's representative in Beijing. As country director for China, I was responsible for business development in this large emerging economic region. My job was to establish partnerships with Chinese banks and companies, and with Sino–foreign joint venture companies, and open up representative offices in major Chinese cities. Through that network I was to find projects that needed funds from the international capital markets.

It turned out to be a fascinating time in history, in which China started to embrace elements of Western capitalism. While the Tiananmen incident in 1989—the crackdown by the Chinese military on massive student protesters who were calling for democracy—was still fresh in the memory, the Chinese leadership had started opening up the economy to foreign trade and investment. It was a time of rapid change and innovation, which gave me a crash course in global capitalism and economics.

Within a few years, the Chinese landscape transformed completely. The majority of the population shifted from wearing Mao suits to wearing Western-style clothes. Traditional houses and Soviet-type apartment blocks gave way to impressive skyscrapers and office buildings, while agricultural land was transformed into industrial zones with manufacturing facilities and four-lane highways that were the products of foreign investment. I vividly remember a meeting with the mayor of Shanghai, who from his office pointed to a huge area of barren land across the River Pu with the words: "That land will become China's new financial center." I could not believe it. It felt like what bankers call a white elephant project—grandiose and unrealistic. But I was proven wrong: within a few years Shanghai converted the swampy wasteland—called Pudong—into a complex of gigantic office buildings and highways, which shortly afterwards became China's financial center, where banks such as ours were only too keen to establish an office.

As China started to report double-digit economic growth, foreign economists no longer spoke of China as a third- or second-world country (i.e., the socialist world), but as an economic powerhouse with global impact. This period represented the beginning of the ascent of China's economic power that continues today. In a sense, it was another Chinese revolution, but this time one that would not lead to political chaos and social unrest. It was a relatively peaceful revolution from which the world (the third, second, and first) was set to benefit, or so it seemed.

1.5 Business in Tibet

The economic liberalization of China allowed me to travel widely and freely; I opened up offices for the bank in Shanghai and other major Chinese cities. The increasing influx of foreign companies, capital, and technology into China, coupled with Chinese banks and firms discovering international markets, meant my business was very busy, but in a sense also easy. As everybody needed advice and capital, the only thing we needed to do was to position ourselves cleverly, as a trustworthy international financier. Since ABN AMRO Bank had grown to become one of Europe's leading banks with a strong capital base, we could pick the best deals. We became particularly successful in funding technology transfers from foreign to Chinese companies. Foreign multinational companies who started to set up production plants in Chinese special economic zones also knocked on my door for funding and advice. My workload exploded.

Since sitting behind a desk full of paper in Beijing's first skyscraper, or at a banquet table with Chinese officials, were not always my favorite aspects of the work, I tried to get out of the office and see the country as much as

I could. Much of China was still clouded in the mystery of isolation, tradition, and poverty. In 1992, the Bank of China invited a delegation of foreign bankers to Tibet, to advise local authorities on how to attract foreign capital and technology. Recalling my inspiring conversation with the Dalai Lama a decade earlier, I jumped at the opportunity. Our group of ten bankers from the U.S.A., Europe, and Japan was said to be the first foreign business delegation to visit Lhasa in modern history.

The journey was a fascinating experience; visits with chain-smoking communist party officials were alternated with excursions to temples, palaces, and farms. Our hosts imposed a strict regime on the foreign guests: the only time we were alone was in our room in the Lhasa Holiday Inn. One evening I managed to escape our host's attention to visit a monastery that was full of devoted pilgrims, some of whom had prostrated their way to Lhasa over a distance of some 2,000 km. Though somewhat physically disheveled, I preferred their company to the artificially arranged visits to Chinese-controlled factories. Of the trip, I most vividly remember the shock when, immediately after visiting the majestic Potala Palace in Lhasa (the traditional "home" of the Dalai Lama), we were driven to the designated economic zone of Lhasa, a rocky no-man's-land. Pointing in the direction of a few sheep that occupied the dusty plain, the government official announced proudly: "This will be the high-tech zone of Tibet. In a few years, it will be just like the bustling trade zone of Shenzhen near Hong Kong." Outwardly polite to our Chinese hosts, our group of hard-nosed financiers ridiculed the plan. Which foreign company would outsource its business to the landlocked roof of the world? Who would be attracted to a marginal and unskilled workforce comprised largely of nomads and monks with no purchasing power?

Yet the fact that these plans were central to Chinese new official policy on Tibet triggered a whole new scenario in my mind. Tibetans may have suffered significantly from political upheaval in the last few decades, but the opening up of their economy could eventually have an even bigger impact on the country and its culture, especially in view of the fact that more than a billion Chinese would-be entrepreneurs are on Tibet's doorstep. It would deeply challenge the traditional Tibetan way of life, even more deeply than communism had over the preceding decades. But it would also, undeniably, offer opportunities for Tibetans to improve their living standards and regain some of their freedoms. The economic reforms in China proper, which had been implemented some years earlier, had also provided very welcome opportunities for entrepreneurial Chinese citizens. The reforms had freed them, to some degree, from political oppression.

Because of my interest in creating a model that generates both economic and social wealth, I started to reflect on the ideal Tibetan economy. As a temporary escape from the busy, booming, and polluted coastal regions, I traveled to other parts of the Tibetan plateau, inland China, and Central Asia to study the effects of market reform. The postcommunist economy of Mongolia—a country with a similar culture and geography to Tibet—was particularly interesting. Recently freed from decades of domination by the Soviet Union, it seemed as if the Mongolians were able to strike a balance between free-market capitalism and the preservation of their culture. I asked my superiors to add Mongolia to my territory for business development, which they granted. So I became ABN AMRO's first representative in Mongolia.

1.6 Reconnecting with the Dalai Lama

Some time later, I was invited to Mumbai to speak at a forum of Indian businesspeople on China's economic reforms. India was about to open up to foreign investment and they wanted to know if China's model was worth following. After my speech, in which I encouraged the Indians to follow, in general, the Chinese market reform approach, I decided to take some time off to travel to Dharamsala and reconnect with the Dalai Lama. A few days later, having flown to New Delhi and driven a long distance through the Himalayan foothills, I was back in the Dalai Lama's living-room after ten years. I was joyful and excited to speak with him again, and less nervous now that I had seen more of the world, including his homeland. Unlike the first time we met, His Holiness did not give me much time to formulate my questions. After the customary exchange of *khatas* (white scarves symbolizing respect) he started to pick my brain on everything I knew about China. He was deeply intrigued about what I had to say about China's economic reforms, its emerging private sector and its impact on the Tibetan plateau.

The Dalai Lama did not apparently object to liberalizing China's or Tibet's economy. "I don't think that economic development *per se* is necessarily a threat to the culture and spirituality of Tibet if, in its implementation, it takes into account the preexisting conditions of the country. Economic development may coincide with cultural development. When we speak of happiness in Buddhism, this also implies material well-being," he made clear.

His concerns were of a political nature: "I strongly support economic development in China. It is a good thing. It has given people better lives and more self-confidence. But I am concerned about the political circumstances of the growing economy. The Chinese people continue to live under a totalitarian regime. Moreover, China has nuclear weapons. If the economy keeps growing under these circumstances, I think we might see some severe consequences, not only for bordering countries like Tibet, but also for a large country like India and ultimately the entire planet."

He paused for a long time before continuing: "Now you see, as far as Tibet is concerned, China's economic development raises serious concerns due to the large number of Chinese that will move into Tibet. Tibet has no authority to keep them out. The greatest threat to Tibet is mass Chinese immigration and subsequent removal of opportunities for the indigenous people, not the development of the economy."

As I had understood that the Dalai Lama was in favor of economic development, at the end of the conversation I suggested that perhaps some sort of investment fund for Tibetans could be set up. I suggested that, with the assistance of foreign people such as myself, an investment fund could be a way of ensuring that Tibetans would benefit from the economic opportunities that would arise and could then compete effectively with Chinese immigrants. The Dalai Lama responded by saying that he felt whatever he did in his name would be considered suspect by the Chinese authorities, so he could not do anything directly. However, he encouraged me to come up with more concrete proposals. He also encouraged me to keep looking at the developments in other Central Asian regions, which had become free from Soviet control and started implementing free-market economics.

Following our meeting, the Dalai Lama arranged for me to talk with key Tibetan leaders in Dharamsala, but I encountered considerable skepticism during these conversations: how could Tibetan refugees ever assist the Tibetans inside Tibet through business? Some even took the view that any business within Chinese structures was only to the advantage of the Chinese and that the only way to help Tibet was through a political solution granting Tibet more autonomy. I was disappointed. However, I realized that many Tibetans had experienced trauma at the hands of Chinese communists, and this had dampened their hopes for a positive outcome.

I remained hopeful that the market-oriented trend in Asia would ultimately be beneficial for the culture-rich indigenous inhabitants of the Tibetan plateau. This could be possible if an appropriate development model could be applied and Tibetans were granted fair opportunity in relation to the influx of Chinese immigrants. Making Tibet an export-processing zone like Shenzhen and Shanghai on China's east coast was obviously not appropriate given the plateau's landlocked position and lack of infrastructure. Clearly, the energy- and capital-intensive industrial model of the West, which was by now also running at full throttle in Asia, would lead to ecological disaster on the dry and fragile Tibetan plateau with its rugged mountains, pristine lakes, and untouched wildlife. Even though the spectacular rise of the Asian Tiger economies was commonly applauded, I was continually reminded of the unpaid price in the form of air pollution. During the gloomy winter months in Beijing, I found a thumb-thick layer of coal dust on my car roof every morning.

1.7 Sustainable development in Mongolia

Following the first global conference on the environment in Rio de Janeiro in 1992, the governments of China and Mongolia—countries under my banking jurisdiction—started to talk about "sustainable development," a mode of development aimed at maintaining natural resources for future generations, and increasing usage of renewable energy. This trend enabled me to shift part of my professional focus from financing large multinationals, power plants, and airplanes, to supporting wind and solar energy generators, water treatment plants, and microfinance institutions (which offered small loans to economically deprived people). Inspired by the Dalai Lama's suggestions, I initiated meetings on sustainable development with Chinese, Mongolian, and Central Asian organizations and started a development fund for inner Asian regions. In response to the warm reception that we received for these initiatives, together with the Mongolian scholar Ts. Batbayar and the British Central Asia expert Shirin Akiner, I cofounded the Council for Sustainable Development of Central Asia (CoDoCA).[4]

---

4   CoDoCA organized two major conferences in Central Asia, in Ulan Bator (1994) and Urumqi, China (1998), and smaller seminars in Bishkek, Kyrgyzstan (1999) and Almaty, Kazakhstan (2000). When, after the 9/11 incidents in New York, the U.S.A. initiated the war on terror, many Central Asian borders closed again, preventing new conferences from being held. See Akiner, S., Tideman, S.G., & Hay, J. (Eds.). (1998). *Sustainable Development in Central Asia*. London: Curzon Press.

Interestingly, this led me to my unexpected next meeting with the Dalai Lama. On the first day of the first major CoDoCA conference in Ulan Bator in 1994, where we had invited delegates from all over Central Asia including Tibet, we heard that the Dalai Lama was visiting Mongolia. Some participants went to the airport to greet him. But I felt that, as co-organizer and a banker in China, I should not be seen in public with the Dalai Lama. China had objected to the Dalai Lama traveling to Tibet's neighbor Mongolia, as this could stir up unrest among the Mongolian minorities in China who, like the Tibetans, regarded the Dalai Lama as their spiritual leader. The title Dalai Lama was actually bestowed by the Mongolian leader Altan Khan in 1578 to a lineage of Tibetan spiritual leaders—Dalai, meaning "ocean," referred to his Tibetan honorific name "Ocean of Wisdom." But halfway through our conference, one of my co-hosts said that he had arranged a meeting for me with His Holiness, as a reward for having initiated the conference.

When I visited the Dalai Lama in a former Soviet resort outside Ulan Bator, he was, as always, immediately to the point. He wanted to know the results of the conference, which Tibetans had participated in it, and what our next steps would be. He also insisted that I gave him an update on the Chinese economy and political situation as I saw it, and how that could possibly help the Tibetans. Rather than just talking myself, I wanted to use the opportunity to seek out the Dalai Lama's wisdom on sustainable economics and business, taking Mongolia as an example. Under Soviet communism all land had belonged to the State. While this has eroded private initiative, it had some benefits for the country's large nomadic population, because all land and grazing issues were presettled and determined by the State. Since the State exported cashmere wool across the Soviet world, nomads were assured of a fixed uptake of their animal products. But after the Soviet system fell apart, and Western-style privatization replaced Communist ideology, the land was up for the taking, and widespread disputes arose over ownership of land and grazing rights. The nomads felt they were better off in the city and gave up their traditional lifestyles.

As this was a typical dilemma in the transition toward a modern economy, not unlike the changing fate of the Tibetans in the opening up of the Chinese economy, I asked His Holiness for his view on ownership of private property. Who should own the land? And how does Buddhism look at the issue?

"Generally, when everything is owned by the State, it dampens individual responsibility and private entrepreneurship. It is clear that in the Soviet Union and communist China this system has to change. Market freedom provides people with more opportunities for business, which generally is much preferred. But this should not be taken to an extreme. The state will have to play an important role, for example in assigning land rights and monitoring proper use of the land, because a free market will not do this by itself. This is the middle way approach—the middle between centralized control and market freedom. This is in line with Buddhist principles. Especially with regard to retaining the traditional culture of Mongolia, which represents strong values such as sharing, collaboration, and respect for nature, it is important that the State guides the process of market liberalization."

To hear His Holiness speak of the "middle way" in the context of economics reminded me of the notion of the "mixed economy," in which the State and the market, or the public welfare and private gain, are rightfully balanced. It transcended the endless dispute between socialists and capitalists on the role of the State versus the market. Both are needed. But in the years after the fall of the Berlin Wall, the world was in a rush to embrace a more liberal version of economics: it was the birth of the neoliberal ideology in which the role of the State was limited to providing the conditions for the market to operate freely. Would Mongolia be an exception? The Dalai Lama continued with a passionate plea to help Mongolia to both preserve and revive its Buddhist heritage—suppressed by 70 years of Communist rule—not just for the sake of Mongolia itself but also for the world at large.

"Mongolia is traditionally a Buddhist country with a culture soaked in Buddhist values, just like Tibet. But contrary to Tibet, Mongolia is now a sovereign country that can freely conduct its own affairs. There is a risk that Mongolia blindly copies free-market capitalism from the West. As I said, this may not work well for all parts of Mongolian society, such as the nomadic population. Mongolia has the opportunity to design its own economy taking principles of Buddhism and traditional ways of life into account, turning the country into a role model for sustainable development and balancing State guidance and free markets properly. Nomads traditionally lived in harmony with nature, because they were so heavily dependent on it. If Mongolia seizes this opportunity, it could be a role model for the development of Tibet, as well as for many formerly socialist countries."

When I left the meeting, I had to work my way through a large group of Mongolian pilgrims, who were pushing hard to see a glimpse of the Buddhist leader; some were even on the verge of fighting to be in front. It was a fascinating display of devotional competition, unlike anything I had seen in other religious circles. What did this say about the prospect of a Buddhist-based economy in Mongolia?

1.8 Roller-coaster banking

This vision of Mongolia as role model and custodian of Buddhist values was a repeated theme of the Dalai Lama. It inspired me to continue with my work on sustainable development in this unique part of the world, where ancient cultures grappled with both the opportunities and the challenges presented by the new postcommunist world. I was glad that, next to my well-paid job as a banker, I could do something useful for people such as the Tibetans and Mongolians, whose culture and religion I increasingly began to appreciate. As China's culture became obsessively materialistic and money-oriented, and economic growth continued at double-digit rates, in contrast the Tibetans and Mongolians cherished a spiritual outlook, looking at life with a sense of amusement, compassion, and devotion. I started to envision the life of a responsible banker who, like Robin Hood, would use some of the wealth generated from the rich (Chinese) for the poor (Tibetans and Mongolians). There were plenty of opportunities to make good money from China's booming economy and so create a "middle way" of living for myself. My fundraising efforts for projects in Tibet and Mongolia flourished: we supported a few dozen schools, clinics, and monasteries, benefiting hundreds of kids and disadvantaged communities. This gave my life a real sense of purpose and meaning, creating value for others and myself.

All seemed well. But on a deeper level, something did not feel quite right. It took a global financial crisis for me to realize what this was.

After five years in China, the bank had suggested that I spend a few years in the Netherlands at the bank's headquarters, so that I could gain new knowledge needed for my career within the bank, before moving abroad again. Since I now had three small children and the quality of life in Beijing was rapidly deteriorating due to pollution, my wife and I welcomed a break in our home country. By 1997, I found myself at ABN AMRO Bank head office in a job overseeing loans and investments in Asian countries. It gave me a new perspective on the entire Asian region, where most countries had now embraced the model of free-market capitalism. A new generation of Asian Tigers had arisen, attracting large amounts of investment capital from the West. No longer did we speak of developing countries: "emerging markets" had become the preferred term.

In the years leading up to 1997, enormous short-term foreign funds flowed into a variety of longer-term investment opportunities in Asia. For example, after more than a decade of heavy foreign investment inflow, Thailand had a booming economy with double-digit GDP and income-per-capita growth.

Bangkok had become awash with foreign goods and high-rise office buildings. Conventional wisdom held this to be a sign of economic health: such a modern city was a sign of affluence and progress. That was until 1997, when economic observers noticed that many of the buildings were empty, and took this as an indication of fundamental economic risk. Bangkok's real estate had been financed by foreign dollars that one day needed to be repaid. The low office occupancy suggested that Thailand would be short of money to repay these debts. This fear triggered foreign investors to sell their Bhat, the local currency, and convert them back into dollars—leading to a drastic drop in the value of the Bhat. The Thai government could do nothing to limit this conversion, for it had fully subscribed to the ideology of neoliberal economics which mandates an open capital market so as to be attractive to foreign investment money. What Thailand's financial establishment (and most foreign bankers) had considered foreign investments aimed at growing with the Thai emerging economy, turned out to be speculative short-term funds which were called back by their foreign owners without warning. They were concerned that their foreign dollars would disappear in unproductive real estate. Many of the Thai banks and companies were overexposed to the dollar. They could not repay dollar-denominated loans in worthless Bhat and went bankrupt overnight, leading to massive lay-offs. The crisis spread to all Asian economies. It shocked the main financial centers of the world.

This crisis could be explained as a normal response to risk in international finance—a "market correction"—and most economists did just that. But my own role in it taught me a deeper lesson. As risk manager for a foreign bank, the moment the first panic erupted, I was told to immediately dispose of Asian assets, regardless of the relationship we had built up with Asian clients over the previous decade. Gone was the bullish language on emerging markets, liberalizing economies, and free trade. My superiors now talked about irresponsible and corrupt Asian governments with cowboy-like bankers and bonanza economies. We went overnight from faith to fear, and from boom to bust. Had we fooled ourselves for an entire decade? I had been proud of working with highly intelligent top bankers, but I could see little cleverness in this sudden shift of mood from hope to despair.

The irony was that banks like ours did not need to worry: we took refuge in the international community and asked the IMF for a bailout, which they generously granted by refunding our nonperforming loans: this "market irregularity" was not our fault. But banks in Thailand, Indonesia, and South Korea had no such escape. They were actually blamed for the currency mismatch. In fact, after the storm had settled, many of them turned out to be cheap buys for foreign banks. The crisis had eroded their equity so they were easy prey for foreign capital.

Another thing that struck me was the fact that China escaped the financial crises without any real harm. China had actually resisted pressure in the 1990s from we bankers to make their currency freely convertible with foreign currency. This non-convertibility, which was common in socialist markets, trumped the mantra of free trade that had been repeated all over the world since the demise of the Soviet Union in 1989. Even though investors panicked over their exposure to China, the currency restrictions prevented them from withdrawing their money. As a result, and to the pleasant surprise of the international investment community, China's economy remained stable. The international financiers took back their rhetoric on market liberalization and instead applauded the Chinese for their prudent governance and far-sighted risk management.

Luckily for the global investment community, Asia's underlying economies remained productive so that, after a year or two, they all regained the faith of investors. Asian workers were less lucky, as many had lost their jobs and houses. The more fortunate ones were quickly reabsorbed into the economy. But the financial crisis had opened my eyes to something that worried me deeply: the extreme volatility of the global monetary system. Perhaps even more significantly, it showed the hypocrisy of much of the international investment community. When we preach free trade and open capital markets to countries in the South and East, we—representatives of the Western business world—are actually saying that we want them to be free for us, not for them. Free-trade capitalism is driven by selfish needs, not by a willingness to share risks in developing each other's economy. And these needs are needs for capital growth. Contrary to what the economics textbooks tell you at school, markets are not merely free exchanges between suppliers and consumers; there is a third player involved with a powerful voice. This increasingly powerful third player is the international financial community, the banks and investors, who have a stake in luring both suppliers and buyers to use more and more capital. The force of the capital market is like a tornado sweeping the world in search of growth, existing due to its own increasing speed and usurping anything that is too weak to resist it. It was thrilling to be a rider on this storm, but I started to wonder where it would all end up. I was riding a tiger that seemed untamable.

1.9 The invisible heart of the market?

The crisis caused me to lose faith in the "invisible hand" of the market, as the founding father of modern economics Adam Smith had called it. There was no hand, and the only thing that was visible to me was an obsession with growth and consumption. We were conveniently ignoring indicators of increasing ecological and social distress. Asian Tigers were revealed to eat up their own rainforests, ecosystems, and traditional cultures. For how long could this continue to work? The whole direction of free-market economics seemed off-target. It dawned on me that the ideology of the industrial society, driven by notions about monetary growth, ever-rising standards of living, and faith in yet another technical fix, is unworkable in the long run. The supremacy of the market, in whose name otherwise good governments often ignore responsibility for the common good, is merely a very fashionable delusion that needs to be sustained for its own sake. Markets are fundamentally flawed: rarely do we account for costs correctly. This has led to stress in the capital markets and a failure to assess the downside of economic growth.

In short, it was as if I was cleaning the deck of a supertanker that was on course for hitting an iceberg. Combating symptoms, while good in themselves, would no longer be enough. It was time to reassess the way we look at the world around us, and above all at how we behave in the pursuit of economic value. Instead of relying on the invisible hand of the market, we should become aware of the heart of the market: people and their environment. There is nothing wrong with economic wealth, but how can we create social and environmental wealth at the same time? The questions I had asked myself as a student were now becoming a quest for understanding: how can we design an economy that works for everyone?

The quest became more personal as the heat was turned up within my bank. The ups (and manageable downs) in emerging markets had increased the appetite of international bankers to look for other growth opportunities, particularly when it was possible to outsource the risks to others. This coincided with the rise of online companies as the next investment targets. European banks started to remodel themselves on American investment banks so that they could participate in the next stage of roller-coaster capitalism. The New Economy—as it was called back then—symbolized the end of traditional risk management. It was no longer necessary to be held back by traditional concepts of (slow and cyclical) growth and decline in the real economy: now, we could all jumpstart eternal growth through the boundless opportunities presented by internet technology and globalization. The sky was the limit. What this New Economy needed was financial engineering: products that allowed the gains to be maximized and the risks to be minimized.

1.10 Maximizing money or meaning?

My superiors told me to shift gears in my career. Up to now they had allowed me to travel widely, establish and maintain relationships in China and beyond, and play my part in responsible banking. But now the message was: I should specialize my work in high-yield financial products and give up fantasies about sustainable banking.

The emphasis was on maximizing short-term shareholder value. Gone were the days when you could invest in long-term relationships without immediate payback in the form of profit; gone too, it seemed, were the prospects for my career in sustainable finance. However, I was not yet ready to give up the life of a banker. I enjoyed the possibilities that it gave me, including the handsome salary that allowed my family a good life in the leafy Amsterdam suburbs, where my kids could cycle to school and play outdoors. But my mind was restless.

I decided to seek guidance from people I regarded as wise, on the more fundamental dilemmas that we as an industrialized and increasingly growth-obsessed society were facing, which I felt were also reflected in my career. So I turned to the Dalai Lama again. This time I was determined to explore these issues more deeply with His Holiness.

But where to start? The Dalai Lama was not trained in political economics or corporate finance, nor had he run a business. I remembered a talk he had given several years earlier on the nature of the human mind. The Dalai Lama said that the human mind is essentially kind, compassionate, and caring. Selfishness and other negative states of mind are like clouds that obscure the fundamentally clear and open nature of the sky. They are not fundamental to the mind's nature. It is possible—and advisable if we want to attain happiness—to remove selfishness by spiritual practice.

Yet free-market economics (and, in the wake of its success, also modern Western culture) was based on an assumption that the common good is best served by allowing individuals to pursue their self-interest. Competition among fellow humans was believed to be not only inevitable but also necessary to bring out the best in us. This assumption had permeated popular Western thinking: compassion is a nice-to-have, not a must-have. Competition ruled as it was thought to lead to optimal efficiency in production, thus optimal consumption, which was regarded as the main purpose of economics. Other values, such as collaboration, teamwork, and caring for others and the environment, were seen as secondary in the pursuit of efficiency, and not a goal in their own right.

So it looked as though these views were diametrically opposed: compassion versus competition, efficiency versus collaboration, and so on. This did not make sense to me. I wanted to understand if we were essentially compassionate or competitive beings. If the latter, I would forget about trying to reconcile "finding meaning" with "making money." If the former, I would take my personal development a bit more seriously, trying to cultivate my supposedly kind nature while integrating this into the business world. By looking at this fundamental question, I felt that we would be able to gain insight into how we can redesign economic systems so that they align with the natural human condition. These insights would allow us to start creating sustainable economics, sustainable business, and sustainable workplaces, and ultimately a sustainable society.

1.11 Compassion or competition?

With this in mind, together with Buddhist groups in Holland, we invited His Holiness to visit the Netherlands and attend a forum to speak on these issues. I invited two friends to be part of the team organizing the visit: Florens van Canstein, with whom I had traveled to India 17 years earlier and who was now, like me, a banker, and Marcello Palazzi, an Italian-born businessman who was a pioneer in the field of social enterprise. By 1999, the neoclassical ideology of *laissez-faire* economics still reigned supreme. Yet critics of the system were becoming more visible; we sounded some of them out. We did not want a black-and-white debate over capitalism versus socialism, or big business versus environmentalism. We wanted to transcend these dualistic notions, explore root causes to this divide, and come up with strategies for the future.

First, we needed to persuade the office of the Dalai Lama. His secretariat had some initial hesitation about adding a meeting with businesspeople to his already overpacked agenda: "If people are interested in what the Dalai

Lama has to say, they can go to his open lectures and teachings. Why should he make an exception for businesspeople?" Since the Dalai Lama's popularity as a man of peace was reaching pop-star-like status, many Westerners wanted to be seen to be close to him. But after I managed to convince his office that we had no intention of using the Dalai Lama as a marketing tool for business purposes, they were quick to confirm the Dalai Lama's attendance. We then looked for other forum speakers who, on the one hand would be appreciative of the message of a spiritual leader like the Dalai Lama, and on the other, were grounded in the profit-focused reality of business. Thanks to Ruud Lubbers, former Dutch premier, we secured the interest of the Dutch government and the Dutch royal family. We received sponsorship from the accountancy firm PwC (whose chairman, Jermyn Brooks, joined the panel), Deutsche Bank, and a number of private entrepreneurs.

On the evening of October 18, 1999, 400 people gathered in the Nieuwe Kerk in Amsterdam to attend the forum with the Dalai Lama, which we had titled "Enterprise and Development in the 21st Century: Compassion or Competition?" A group of distinguished representatives from European business, government, and academic circles was invited to exchange views with the Dalai Lama. The Crown Prince of Orange, Willem Alexander, was among the audience alongside leaders from business, government, and the not-for-profit sector.

This unusual group of people came together in an equally unusual setting. The Nieuwe Kerk is the traditional venue for the coronations of Dutch royalty. The former church building was displaying a wonderful exhibition, "The Dancing Demons of Mongolia," which I found auspicious given my meeting with the Dalai Lama in Mongolia some years before. It featured stunning Buddhist relics and artifacts from Mongolia, along with Mongolian musicians playing live before and after the forum. The setting lent itself to an atmosphere of reflection and appreciation.

The topic of the forum was no less unique, for religious leaders and the business community rarely meet to discuss the role of enterprise and economics in our societies. Finally, the timing of the event—on the eve of the 20th century—was extraordinary. Now, almost two decades later, we cannot help but feel that the forum was extremely well timed. It focused our thoughts on how to deal with the choices that were ahead of us in the new millennium. What do we consider natural and desirable? Cooperation or conflict, dialogue or violence, sustainability or the bottom line, compassion or competition?

The Dalai Lama at the "Compassion or Competition" dialogue in 1999.

2
First dialogue: Compassion or Competition (Amsterdam, 1999)

The first dialogue focuses on the overall trend of the social role and impact of business. The Dalai Lama and his dialogue partners—Hazel Henderson, Geoff Mulgan, Jeremy Brooks, and Ruud Lubbers—each present key themes that they think will affect business leadership in the 21st century. This is followed by questions and comments from a selected group of leaders from business and society and a few questions from the audience. The reflections by the Dalai Lama inspire the following key points:

- The changing reality for business, such as the increasing gap between rich and poor and environmental degradation, requires a new type of thinking in business, one that extends the usual narrow perspective on immediate results to a wider, more holistic, view that takes indirect impacts into consideration.

- There is no conflict between business and ethics. Ethical conduct in business is necessary for achieving long-term benefits. Without it, business may benefit in the short term but, in the long term, it will suffer enhanced risks, loss of reputation, and lack of trust among customers, investors, and employees.

- The importance of motivation for people working in business—a compassionate attitude will benefit others and oneself, including one's business.

- There is a need to change the rules of the game in economics and business. It starts with personal leadership, which can change organizations and eventually the larger system. This engenders a process toward "compassionate economics" and "responsible markets."

Sander Tideman: The values of compassion, tolerance, and wisdom, which the Dalai Lama so convincingly and humanely advocates, should not be confined to temples, churches, and religious circles alone; rather, those values should also be underpinning the world of business, economics, and finance, which controls so much of our society today. While most countries in the world now practice free-market capitalism, it is clear that the market alone cannot solve growing social inequality, poverty, and environmental degradation. In running the increasingly complex, interconnected, and volatile global economy, we need to be guided by a vision of well-being for everyone on Earth. This forum is an attempt to align the world of enterprise and economic development with human values and ethics.

Geoff Mulgan: We are here to talk about as important a set of issues as one could imagine in the last weeks of the millennium; above all, whether the next period will be one of ethical progress or not. Some 70 years ago, when Mahatma Gandhi was asked by the king of England what he thought of Western civilization, he said, "It would be a good idea." And it's worth remembering that this was the time when Europe's empires ruled the world, the Holocaust was being prepared for, and the gulags were at their peak— which is a sign, perhaps, that *some* ethical progress is possible. But since then, we've also seen environmental destruction and, in more recent times, genocide in Bosnia and East Timor as bad as anything in human history, which is a reminder that progress is by no means inevitable.

We are here tonight to talk particularly about the role of business, which is quite rare in discussions of this kind, but quite appropriate given that more than half of the world's largest economies are now businesses, and not nation-states.

And yet, I think it is fair to say that businesses have found it hard to work out what this implies for their responsibilities, whether in relation to human rights, child labor, the environment, or new technologies such as genetically modified foods. And businesses are taking part in the wider debate, which His Holiness has been a key part of, which is about how we rethink our responsibilities and catch up with the new realities of interdependence.

The Dalai Lama: It is a great pleasure and honor for me to participate in this brief discussion on the role of compassion and competition in modern society. As far as an informed opinion on the role of business in modern society—specifically the ethical responsibilities attached to doing business—is concerned, I don't think I really have much to offer. But in any case, I think this kind of gathering, this kind of serious discussion, is an excellent opportunity to learn; so I'm looking forward to listening, rather than talking!

It's my fundamental belief that, generally speaking, every human being has a moral responsibility toward humanity, a responsibility to consider our common future. I also believe that every individual has the potential to make at least some contribution to the happiness and welfare of humanity. Certainly those of you who are in the business world, which is an important part of humanity, have great potential to make a contribution. Therefore, it's important that, while you are engaged in business or money matters, one part of your mind retains a sense of this responsibility. If you think only of the immediate profit, then you will have to suffer [negative] consequences in the long term. I think that's evident from what's happening to the environment. The results of the consumption of natural resources, and of manufacturing regardless of the impact on the environment, are now clearly appearing. So, those of you in business have a certain connection with this situation, and bear some responsibility.

Moreover, at the global level, there's this gap, this huge gap, between rich and poor: the Northern industrialized nations enjoy a surplus—in America, for example, the number of billionaires is increasing rapidly—while at the same time, in the Southern hemisphere of this very same planet, fellow human beings in their thousands and thousands lack even basic necessities, and their children display clear signs of the malnutrition that will blight their whole lives. This is very sad. And within countries too, the rich increase their wealth, while the poor remain poor, and in some cases become even poorer. So, that reality is not only morally wrong, but also a source of practical problems. Even though the constitution may provide for equal rights and opportunities, the tremendous economic inequality places those poorer people at a disadvantage in terms of their education and careers. This leaves them with feelings of lingering discontent, and a sense of impoverishment and inferiority. Those feelings transform into hatred, which in turn leads to crime, gang culture, murders, and so on. As a result, the whole of society suffers—even the wealthier people, who are constantly under threat and plagued by feelings of insecurity. And I think at a global level too, unnecessary problems sometimes arise because of this big gap. So certainly the business community is clearly connected with these problems as well.

In conclusion, I believe that any human activity carried out with human feeling, a sense of responsibility, a sense of commitment, a sense of discipline, and a wider vision of consequences and connections—whether it be involved with religion, politics, business, law, medicine, science, or technology—is constructive. On the other hand, if these human activities are carried out with short-sightedness and for short-term interests, especially if the intention is simply to accrue money or power, then they all become negative, destructive activities. If your mental attitude is not right then even preaching religion is destructive and creates more trouble. So everything depends on the human motivation. That's why, I believe, in order for any human activity to be constructive, one must first check one's motivation.

And when we are talking about motivation, I consider the most important aspect of motivation to be a sense of caring for one another, a sense of sharing with one another, and a sense of responsibility for the big issues related to the common interest.

So, that's my view on how to make money *properly*!

2.1 The rise of socially responsible business

Hazel Henderson: I want to thank the Dalai Lama, His Holiness. I hope this will encourage members of other religious traditions to speak out on the subject of the economy as well, because, of course, it's not something separate from our lives.

My concern really goes beyond current globalization, to the question of how we shape a sustainable global economy. To me, this question is about four Cs (at least in English): competition, cooperation, compassion, and creativity. And so we have design work to do: to change the shape of our economy, to make it ecologically sustainable and socially sustainable. And the key to making it socially sustainable must be to reverse the growing gap between the rich and the poor, not only between the North and the South, but also within many of our countries, certainly within my own country, the United States, where that gap is getting wider. So I very much support the campaign that's been going on called "Jubilee 2000." It's very exciting to me that all of these people from around the world actually persuaded the G7 leaders in Washington, including President Clinton, to support the idea of forgiving the debts of the 36 most indebted countries in the world. This is necessary—but not sufficient.

We must also change the current recipe for economic growth—known as the "Washington Consensus"—because it has steered our countries in the wrong direction, measuring growth by the growth of GNP [gross national product]. Deregulation, privatization, free trade: it's not enough, it's been failing and leading to this widening gap. The problem is that we've focused on very narrow kinds of globalization: globalization of markets, finance, and technology. This has created the financial bubble that I've been talking about for many years, this US$1.5 trillion which goes around the world every day. It's very interesting to me that even one of the economists who in the past ten years have advocated this kind of Washington Consensus, Harvard University's Jeffrey Sachs, has now changed his mind. I was just with him in Prague with President Havel, and he now supports Jubilee 2000, so we are making a little progress.

What we need to do, I believe, is go toward full-cost prices for products, prices which include all the environmental and social costs and, of course, change the formulation of the GNP to reflect a much broader quality of life. I think we also have to shift our taxes away from taxing incomes and payrolls, and instead tax waste, pollution, and the depletion of resources, because this will help to create full employment while improving the environment.

I think that what I have really been doing is trying to raise the ethical strata on the global playing field, because I have been involved in socially responsible investing in the United States since 1982, when it was a "new idea," and we—the groups that had these ethical, mutual funds—were very small. Today, it's very exciting because there are now 60 funds in the United States that copy our funds, and we're very happy about that. And US$1.3 trillion are invested in these funds that don't invest in companies that manufacture weapons, or pollute the environment, or are unfair to workers, and all of these kinds of criteria. I think that we can encourage more corporations to sign these codes of ethical conduct, and develop these ideas of reporting social and environmental costs in their annual reports.

It's also encouraging to see how many people are now involved in the microcredit revolution to make credit available to the smallest village enterprises. Even the big banks are getting involved in this. We have had our own stock market indicator of our socially responsible investment funds, called the "Domini Social 400," in the United States. It's very nice to see that this index has outperformed the Standard & Poor's index every year for the past five years, so that today even Dow Jones has developed a sustainability index of its own. I'm not too sure what they mean by it, but at least they are moving in the right direction.

So I think we really can make quicker progress, if we encourage all of the activities of civil society and work together to encourage corporations to fulfill these kinds of social responsibilities.

2.2 Toward the triple bottom line

Jermyn Brooks: I'm very tempted to take up the challenge—which was not in my script—to respond to the comments His Holiness made about the wealth gap, because this truly is, longer term, one of the biggest challenges, and I'd like to talk about how business could respond to that and make a real contribution.

I'd like to go into the area of the way business is moving in terms of its attitude to business ethics. Now, historically, it was only yesterday that Milton Friedman said, "the business of business is business."[1] And I'd like to show you how we've moved on—at least in terms of some of the more enlightened companies—from that statement.

However, it is very important that we start out by acknowledging that we've not found a more effective way of ensuring the efficiency of business other than by allowing competition between businesses; admittedly, modified by antitrust regulations, and of course with a social safety net for the socially deprived and less fortunate. Nevertheless, on a large scale competition is vital, and it remains the aim of every business person to operate profitably within that competitive environment, and if they forget that, the business will go under. So let's remember that this must be a prime concern for all businesspeople. And of course, this has led to concepts that are very familiar to most of us, such as maximizing shareholder value and—driven very much by the capital markets—the short-term, bottom-line approach to maximizing shareholder value.

What is now happening is that people are beginning to realize that a longer-term approach to shareholder value is bringing dividends, and we heard from the previous speaker, Hazel, that there is at least some evidence

---

1   Friedman, M. (1970, September 13). The social responsibility of business is to increase its profits. *New York Times*, p. SM17.

that companies espousing a longer-term and wider approach to the purpose of business are being at least as successful, if not more successful, than companies with a strict bottom-line focus.

Now, the longer-term focus, in terms of shareholder value, we can put equivalent to a concern about stakeholder value enhancement. What does that mean? It means focusing on the individual components of that stakeholder population, that is, the people who work for the company (that seems so natural, but in many companies it tends to be neglected); it means focusing in a real sense on the needs of the customers, the concerns of the suppliers; it means a concern for the use of resources and the environment; and it means working with local communities, having good relations with the government, and with society at large. A focus on stakeholders also means that companies need to make sure that they are actually adding value, *societal* added value. And what does that mean? It means that if companies, over the long term, are not operating in line with the expectations of society, they will run into major difficulties—and I'll give a couple of examples later on.

So what are the immediate implications for companies taking this longer-term view and focusing on stakeholders? Well, first of all, they need to reevaluate their business values, their ethics, and their strategy in terms of sustainability. It's not sufficient though to go through an exercise of printing nice, new value statements—there has to be a monitoring process. And so a compliance function needs to be set up to make sure that everybody in the company actually understands what the values are, and that there are sanctions and praise for the appropriate behavior. At the same time, many companies have worked on "reputation assurance programs," designed to anticipate the problems you can run into if there are major issues in relation to your reputation. And again, there will be examples in a moment.

The second implication is to open a dialogue with society. Now, that sounds rather grand but what it in fact means is being able to discuss openly, much more than business traditionally has, with NGOs [nongovernmental organizations], with government representatives, with customers and suppliers, and—believe it or not—with your own employees. Many companies train their employees well, hire them well, and, if they're good, even fire them with compassion, dealing with their problems. But do they actually talk to them? Do they use them—as my firm is only just beginning to learn to do—as a source of strength for their own planning?

And the third implication, for me, is that companies need to embrace what is increasingly being called the "triple bottom line" approach to measuring business success, the three elements of which are financial performance, environmental performance, and social performance. The most advanced companies in this area are already preparing management reports covering all three criteria. Particularly in the financial area, some of them are developing forward-looking indices ("value reporting," as opposed to just historical financial reporting), green reports, and social reports. Some of these companies are now publishing these reports, and publishing them in a way that is quite new, admitting publicly where they've fallen short of their own internal standards. And as a final step, just as financial statements are reported on by outsiders, external and independent auditors verify the accuracy of the statements being made in these green reports and social reports.

Now it's rather sad, but I suppose it's a reflection of human nature, that some of the companies with the best records in this area have chosen to adopt these measures as a result of major crises. One of the companies, which is very well known in this country, Anglo-Dutch Shell, ran into both environmental and human rights problems. As many of you know, they faced significant criticism over the disposal of a major oil rig in the North Sea, and, in terms of human rights, through the treatment of a community in Nigeria who were left dispossessed of their homeland as a result of Shell's exploration activities. Shell's response could have been to regard that as an unfortunate PR mishap and to run a program using PR activities and support to try to overcome the negative impact in the newspapers. What they actually did was to go through some of the steps which I've just been outlining to review their values, to introduce compliance systems, to think very hard about sustainability, and to prepare the kinds of report I've been talking about. And in terms of interrelating with society, they've implemented one of the best programs that I've ever heard of via their own website, where they have cross-references to all the NGOs who criticize them, with the opportunity then for anyone who looks up their website to be able to get the story from both sides.

Another example—an American example—is Nike, the running shoes people. They ran into a serious issue when it was discovered that they were employing child labor in sometimes appalling conditions in some of the poorest countries in the world. Their response to these criticisms was to review their whole manufacturing process outside the United States and to think again about what kinds of human rights value they should be adopting in their manufacturing process. They then established a system and asked that it be verified each year in a way that they could share with the press, business analysts, NGOs, and, of course, with their shareholders. One of the most interesting aspects of this case was that Nike didn't necessarily stop employing child labor, nor did they simply close down the factories and deprive families of the income on which they depended for their livelihood. Instead, they entered into a discussion with the people in the countries concerned who could help improve the work conditions for young people and give them opportunities to study and advance themselves.

So, these are some of the examples of the unconventional ways that companies have dealt with problems.

Finally, in summary, I believe that companies can indeed prove that they can operate efficiently, that they can operate profitably, while at the same time complying with the highest ethical values of the societies in which they operate.

2.3 The increasing role of civil society

Ruud Lubbers: I totally agree with the statements made by His Holiness and Jermyn Brooks, and in particular, I agree that there is a new perspective developing in business; businesspeople are internalizing societal values as evidenced by their mission statements, codes of conduct, and so forth. And I'd also add, as a former politician, that it might be useful if society at large— society's leaders, and also NGOs practicing societal values—put some pressure on or continued putting pressure on business, because the only way we can have them [businesses] internalize those values is to apportion blame when they fall short. Jermyn Brooks gave the examples.

And I think this is the future for business: it *will* happen. It's a perspective of hope, maybe you'll call me naïve, but I'm pretty sure that people will be converted to these new standards, this new style of business. Yet we still have a long way to go.

Let me say that in preparing myself for a few remarks here today, I read a wonderful book that records a conversation between the Dalai Lama and Fabien Ouaki,[2] and I took a few points out of it.

The first point was that His Holiness states the importance of the principle of religious pluralism and that it is better for the world to have different religions. In both these statements he makes it clear that religious

---

2   The Dalai Lama, & Ouaki, F. (1999). *Imagine All the People: A Conversation with the Dalai Lama on Money, Politics, and Life as it Could Be.* Boston, MA: Wisdom Publications.

opinion, or religious culture if you like, can be positive under the condition that it cherishes diversity—a plural society. And I think this is very important, because religion has not always encouraged diversity and has been the cause of disputes and war. Nevertheless, religion can fulfill a very positive role when it embraces diversity.

The second point which struck me came at the very beginning of the conversation, where His Holiness states: "Most legal systems refer only to human rights and do not consider the rights of animals and other beings that share the planet with us." I had to think back to 1992, when I had the privilege to be in Rio de Janeiro for the United Nations conference on the environment, and how surprising it was that so many NGOs and representatives of indigenous peoples present made the same plea for a different way of looking at the world. And I realized, as a Dutch person, as a European, that it is indeed more than 500 years since we last heard something similar, in the *Canticle of the Sun* by St. Francis of Assisi.[3] St. Francis was someone still capable of approaching life in terms of brotherhood with animals, seeing the essence of nature and the value of it. Though economically and technologically we have been so very successful in our Western civilization, we have lost the talent to see how we are a part of the Earth, "Mother" Earth, and the universe even. There is something beyond human beings and certainly beyond economics and technology. Since Rio, we have seen the enormous efforts of many people to vocalize this in an "Earth Charter," which is not at odds with the human rights concept, but indeed, adds new elements to it.

It's the paradox of our times that globalization in terms of economics and technology is drawing increased attention to spirituality. It's good, Your Holiness, that you received the Nobel Peace Prize, but it's about more than the traditional concept of peace. You've also been able to send a message to us, just as you're doing here today, about other values. I think this is great. And by the way, it's great that on the very day I see you in this country we hear that this year's Nobel Peace Prize is going to Médecins Sans Frontières [Doctors without Borders]. It is another indication that the concept of peace, the concept of living together, the concept of living in harmony, is very much related to a spiritual approach to life. It's being given shape and substance, not only in politics, but also by religious leaders and by those who prove through their actions that there are other values as well.

3   Doyle, E. (1996). *St. Francis and the Song of Brotherhood and Sisterhood*. New York, NY: Franciscan Institute.

From left to right: Ruud Lubbers, translator Thubten Jinpa, the Dalai Lama, Geoff Mulgan, Hazel Henderson, and Jermyn Brooks.

In Ouaki's book, *Imagine All the People*, we also read about technology. I found this interesting and I'll quote you, Your Holiness, because your ideas are so nicely formulated. At one particular point in the conversation you said: "We had a rule in Tibet that anyone proposing a new invention had to guarantee that it was beneficial, or at least harmless, for at least seven generations of humans before the technological invention could be adopted." And once again I thought back to Rio de Janeiro, where we formulated a so-called precautionary principle, which was very much in line with this idea and—in contrast to the short-term thinking that Jermyn Brooks pointed to as one of the dominant aspects of our modern Western economy—very much intended to promote long-term thinking.

I'd like to give you a report of a conversation I had with Michel Serrer, a famous physician who is a member of the Académie Française. He began by explaining to me that he'd been a professor to generations of students and told me how proud he was of his students and all their new inventions. But then he added something. He said: "Mr. Lubbers, in recent years I've come to the conclusion that more and more of my students, those who I appreciate a great deal, are not at all so proud of being leading physicians inventing and applying new technology. They start to become a little concerned about whether what they are doing is always right. We've had this experience before, you know, the first time with those who brought us the nuclear weapon; they ended up with quite a hangover. We see the same thing happening more and more." I asked Monsieur Serrer, "Do you have any idea what to do about it? Do we, as politicians, have to keep check of our new technology?" "No," he replied, "I've thought about it a lot, and I think we have to go back about 2,500 years to find the answer." I was really surprised—what can we find there, before Christ? And then he explained to me: "At that time there lived a man named Hippocrates. He lived in Greece at a time when substantial advances were being made in terms of what physicians could achieve, and people were becoming a little bit scared about the their power. They discussed the situation and came to the conclusion that what was needed was an oath that would have to be sworn by anyone practicing medicine: the Hippocratic Oath. I think that today we should propose a similar oath for those who are active in science and technology." Indeed, he had already worked out a formula for such an oath.

The point here is simply that we have come so far, to the limits of economics and technology, mastering the whole world along the way, and at the very same time we come to a point where *responsibility*, in moral terms, is more important than ever.

Jermyn Brooks said something along these lines and I applaud that; Hazel Henderson explained that even as famous an economist as Jeffrey Sachs is changing his attitude.

Let me end by stating that I'm really concerned, precisely because I was in business, I was in politics, and I have tried to look objectively at how the world is developing. I'm especially concerned about the so-called Western civilization that the moderator mentioned. Yes, we all preach certain values, but the reality of life is very much what I call the "Bermuda Triangle" of values. What does that mean? I'll sum it up.

First, there's the **economization** of life: it's all about money and the economy. Even human identity is equated with money. Your identity is your income, your level of consumption. Second, there's the enormous power of the media, which has become as short-term in its focus as those involved in the economy. It's all about sound bites. It's all about news that shocks. When we come home we ask, "Has something happened today?" and if nothing has happened we are not satisfied. Journalists know this, so they're constantly adapting the news to meet this appetite. I call this the "**mediazation**" of life. And third, there's **short-term politics**. Politics: it's supposed to be for the public cause and the long term, but in fact it's about doing well in the media. All three of them are about the short term, and about money. If you're in the triangle, you're lost and you sink. Now, we must organize something to counterbalance these effects, and only through adopting spiritual values and a different lifestyle can we achieve this. We must support those who are active in these areas. We must envision a different world and actualize it.

I said a few words about the Earth Charter initiative, but there are a number of such initiatives. Religions of the world have come together, entered into dialogue, and made declarations. What do they propose? No longer being active in politics myself, I have had some time to study these documents. Three elements, reflecting three aspects of society, run throughout: these declarations all aim at a **just**, **sustainable**, and **participatory** society.

Just is a concept known to everyone. It's about fairness, about justice, a concern for all generations. However, we are *very* concerned about this, and His Holiness brought it to our attention once again in relation to the gap in incomes and our capability to do anything about it.

The second element is newer—it's the concept of **sustainability**. At the very moment that we've managed to spread economics and technology all around the globe, we are coming to the conclusion that a short-term approach doesn't work and that we have to take more responsibility for the generations to come. That's exactly what sustainability is about, the long term.

The third is about the **participation** of human beings. Each and every human being has the right and the potential to participate in society. That's really what we want with a just, sustainable, and participatory society. This is about having a job, of course. But it's also the right and the opportunity to be active in other areas, including, yes, contemplation. Therefore, the concept of a more spiritual world and a world in which people can participate is not only about the economy, it's about harmony in life as well.

I think that such a just, sustainable, and participatory society is supported by humankind, by religious leaders, and by NGOs. Practicing those values can counterbalance that "Bermuda Triangle" and that's what we need. Therefore, I was so pleased, Your Holiness, to read your interviews. I've never been in your country, and frankly speaking, to me you are a strange person—maybe a saint, maybe a leader—but faraway and strange. Nevertheless, if you would allow me to say so, I feel like we are partners sharing things, doing things together. I'm grateful you came here to visit us: thank you so much!

2.4 Widening the perspective: business needs ethics

Geoff Mulgan: I'd like to ask His Holiness to respond to the three presentations we've heard. Perhaps I could ask you to address a question that underlies all three, which is: to what degree are we now in an era when good ethics is also good business? And to what extent are there still difficult trade-offs and choices to be made in relation to equality, precautionary principles, and so on? How easy are these questions, in other words?

The Dalai Lama: I have learned many things and I'm grateful. It also gives me some kind of encouragement. Thank you.

In my view, ethics means something *right*, and right means something beneficial to us. So, any action—whether it's business or any human activity—done with that aim and in accordance with a vision of something that benefits humankind and the world at large, is ethical. Of course, sometimes there's contradiction: short-term interests may be bad for the long term and long-term interests may present difficulties in the short term. On the other hand, sometimes short-term and long-term interests coincide. But I think the problem here is due to short-sightedness, as you mentioned, too much exclusive concern with immediate benefit. Also, I think, focusing on a limited area is a problem.

One of the characteristics of the modern Western approach to addressing issues and problems is this tremendous emphasis on clarity and precision. But the problem that arises with an emphasis on precision is that our outlook narrows: precision can only be achieved if the focus is narrow. Looking at the same issue from a wider perspective means one cannot demand the same level of clarity and precision. Similarly, if we extend our perspective to the long-term future, there will be a lesser degree of precision. Consequently, some of our present activities are of some benefit, or at least do not do any harm, but may have negative outcomes in the long run or further afield.

Now the application of ethics here entails acting in accordance with reality, especially the long-term reality. However, I feel that there is sometimes a gap between reality and the prevailing mind-set in business, which tends to be rather

> The problem that arises with an emphasis on precision is that our outlook narrows: precision can only be achieved if the focus is narrow. Looking at the same issue from a wider perspective means one cannot demand the same level of clarity and precision.

narrow. I think today's reality is much changed from the reality of the early part of this century, in terms of population, technology, and also levels of consumption. The world is becoming smaller, too. But our minds tend to lag behind, still caught up in old ways of thinking. So even though we try to solve problems, our efforts are often not so effective, or lead to further, unnecessary, problems.

The main point I wish to make is that when we talk about ethics we should have a clear understanding of what we mean by ethics. In my view, ethics should not be seen as only being embedded in religious faiths, but rather understood more in terms of acting in accordance with the reality of our world. For example, the previous speakers all pointed out how the business world is changing in response to a modern reality, how certain forms of behavior which were acceptable in the past are no longer considered acceptable, and how persistence with these ways of behavior is detrimental even to the financial interests of the business communities. What this suggests is that to act ethically we have to act in response to the reality of the situation, taking into account both the long term and the wider picture.

> Ethics should not be seen as only being embedded in religious faiths, but rather understood more in terms of acting in accordance with the reality of our world.

Geoff Mulgan: I'd like focus a bit more on this question of how our institutions of business and politics, and our mind-sets, need to catch up with the reality of interdependence and of understanding how the consequences of our actions manifest in the longer term. How can compassion in business benefit both society and business?

2.5 The benefits of compassion in business

Henk van Luijk: Your Holiness, in your many lectures and writings you elaborate extensively on the principle of compassion, and you apply it to the realms of education, law, politics, health, and even technology. But also, repeatedly, you admit to having some difficulties in applying it to the domain of economics. If I may paraphrase some of what you have said, you suggest that where competition and profit-seeking is the rule of the game, there is no straight way to compassion.

Now, if this were the last word, it would imply that the dominant fields of human activity would remain outside the influence of compassion. I can hardly believe that we should leave it at that and I'm sure that you agree on this point. However, I also doubt whether we should, and whether we could, turn to businesspeople as individuals and believe they can embed compassion in their work. Many businesspeople are human (to a large extent at least!), but that does not necessarily enable them to permeate, as it were, their field of action with compassion, given the pervasiveness of the laws of the market. To behave in the market in a compassionate way you need allies who admonish you to accept your responsibilities and who are prepared to cooperate with you in a common endeavor. This means that, in the market, the name for compassion is "cooperation" or "responsibility." It is only in this indirect way that compassion can take shape in the domain of economics. I must admit that I feel a little uneasy with this conclusion.

So, here is my twofold question to Your Holiness: do you agree that in market relations, compassion should be interpreted first and foremost as cooperation and responsibility, not simply as a personal trait of character, and, if this is the case, do we, by this translation of compassion into cooperation and responsibility, lose something of the depth, warmth, and power of compassion?

The Dalai Lama: As to the first part of your question, I think the role of cooperation in business, in fact, implies compassion. This is actually quite similar to the idea of how businesses are changing by taking greater care of the needs of the stakeholders, the employees, and so on that one of the speakers discussed. Although the main motivation may be to ensure one's financial success, in the process one is compelled to take into account the needs and concerns of the stakeholders and so on. When you bring about greater cooperation in business, then there is a role for compassion.

Why do we actually need compassion? My answer is, "Because it benefits us." The more compassionate one's mind, the happier one feels. Look at other people: if they are negative and harbor feelings of hate toward others, they lose their own happiness, their own peace of mind and they suffer! So, this is my main argument. I am Buddhist

> Why do we actually need compassion? My answer is, "Because it benefits us."

and I'm practicing compassion, but I do not do so in order to please Buddha. Neither do my Christian brothers and sisters practice compassion merely to please God. No, it's one's very own future that's at stake. Therefore, even if in business your main interest is a successful company, you can still take care of your workers and customers: show them a smile—not an artificial smile, but a genuine smile—then more customers will come!

The second point is a more philosophical question, and I don't know the answer. I'd need to think more about whether reinterpreting compassion in a particular context in terms of cooperation means that it loses its special meaning and power. This is a philosophical question that would require a certain amount of time and thinking. Perhaps I need to do some homework on this, think it over, and discuss with more people.

Erica Terpstra: I have had the privilege of spending the two previous days among 9,000 people gathered in The Hague to hear His Holiness' warm teaching on compassion and to meet the embodiment of compassion and love. I saw young people and older people, I saw businessmen and people without any work, and it was truly heartwarming; I think that this really is the inspiration we need. I asked myself why it is that nowadays Buddhism is flourishing so much in the West and around the whole world. I think that maybe it's because we have too much materialistic greed and too much technology in our modern lives, and maybe, deep inside, we profoundly need to find a new balance between modern life and spirituality. Of course, that not only applies to business but to politics as well.

I remember a couple of years ago I had the privilege of being in a Buddhist monastery for a week of meditation and practice. The day I returned to Holland, I had to go to a political meeting where we debated the so-called half percent more/less. Your Holiness, I'm sorry that I'm unable to explain this to you; it was just about the difficult question of whether the wages of employees should be half a percent less or more. Imagine—I'd come straight from the monastery and was sitting in this meeting, listening in confusion. All of a sudden I said: "Mr. Chairman, I miss some *spirituality*"; they all looked at me as if to say, "Bring out the men in white coats!"

Hazel Henderson talked about the four Cs, Your Holiness, and one of the Cs was competition. Jermyn Brooks said that we should not forget that competition is vital. How do we reconcile competition and wanting to be the best, with the practice of compassion, with the practice of altruism? If you say to people in the domain of economics, "Well, you shouldn't try to be the best because you've got to abide by an ideal of being of benefit to all sentient beings," isn't there the danger that you will be looked upon as a "softie"? Please help us, and teach us.

The Dalai Lama: Usually I make a distinction between two kinds of competition. One kind of competition is more negative: you want to reach the top and because of this you actually create obstacles for others. Alternatively, one simply accepts that just like oneself, others also have the right to reach the top, and if one works hard and determinedly with that attitude, then there's nothing wrong. In the spiritual field also, in Buddhism for example, the aspiration "I want to become Buddha" is not selfish. This is not a matter of wanting to be better than others are, not at all; rather, in order to help others more, in order to serve others more, one needs to have greater ability.

Moreover, competition in terms of the desire to be the best can be applied to many things, not just to profit. One could be the best at bringing the most benefit to people. If a company benefits more people than its competitors, that's really the most important facet of being the best. If a business makes a lot of profit but earns itself a bad name, that's not the best! So I think it's important to have an understanding that the criteria for being the best are not purely monetary. Therefore, I think the desire to be the best is absolutely right. Without that kind of determination there's no initiative, no progress. In the present context, compassion also means the desire to help, to serve, or bring more benefit to the larger community, by any means, including business. So I think that wanting to be the best can go together with compassion.

> If a company benefits more people than its competitors, that's really the most important facet of being the best.

And another point. If compassion were always to mean giving, then any company that acted compassionately would soon go bankrupt! But from the Buddhist viewpoint, what we actually need to do if we want to help others is *empower* them to stand on their own feet. It's not a question of limitlessly giving without any kind of initiative on the part of the recipient. So I feel this idea of empowering others through one's own help could have direct relevance in the business world as well.

2.6 Changing the rules of the game

Eckart Wintzen: Your Holiness, I think you've given us great advice that immediately enters the heart, exactly where it needs to be. You talk about caring, about sharing, and about compassion. But, in this room now, we are talking about competition, business, and Western enterprises. One way or the other it seems that the very fabric of Western enterprise doesn't allow much space for caring, charity, and compassion. As for sharing with the poor, well, we have learned to do that with our employees, more or less. But when it comes to sharing with other parts of the world—what we normally call "development business"—it doesn't usually have much to do with sharing; it's about developing a market. At present, sharing with the next generation seems to be our biggest dilemma, particularly because we don't see the future of the next generation that clearly. You talked about looking at reality, but the reality of the current generation is already a little bit obscure, let alone the reality of future generations. Where things are unclear and don't fit with our present thinking, we have the tendency to hide them, or to dismiss them: "Oh, the problem is not as big as you think, take it easy," and so on.

As I said, it looks as if the Western free-market economy is driven by the forces of *having* and *getting*, which are completely different from the values you're promoting. The Western free-trade economy generates strong egos: *I* want to have the biggest company, *I* want to have the most powerful company, and so on. It generates greed: we want to have *more* employees, *more* assets, and the shareholders demand that. It doesn't generate sharing at all. The rules of the game are wrong in the Western economy—that's our problem.

> The Western free-market economy is driven by the forces of having and getting, which are completely different from the values you're promoting.

Even though it seems to be so difficult to change the rules of the game, as Hazel has already said, we must try to change the rules in such a way that we achieve greater sharing. We need different definitions, we need a different economy, and we need different taxes. Taxes are all about sharing: more of a community's money going to the poor. So we need different rules there. I loved what Ruud Lubbers said, quoting you, Your Holiness, about requiring inventions to be beneficial for at least seven generations—that's great! I think most of our inventions today have damaging outcomes even within the following year and, if not, definitely for the next generation. For example, we have learned to deal with pollution to a certain extent, but we seem to hide away the problem we are creating for future generations by combusting fossil fuels and producing enormous amounts of carbon dioxide.

So we need different rules. I think maybe the onus is on us here in the West to change our politics, our business lives, and our society. But we also need a different attitude at heart. We need change at the level of the individual, the man at the top who is (sometimes!) a human being. The person at the top is always the "best," but very often not in the way that you define the term, since the rules of promotion are not the same. So, we need more decisions coming from the heart, more decisions made with real vision. We all love listening to you sharing your values with us, but we need more of it; give us more, I would like to say, to help us change and look to the future.

The Dalai Lama: I agree very much that fundamental change really needs to start from the individual. Once transformation takes place within an individual, however limited the effects may be, at least within the limited domain of his or her activity, the influence will be felt.

Hans Opschoor: It's not so much a question that I have but more of a statement. Maybe that gives the Dalai Lama a little rest! Indeed, one of the great challenges is fairness across generations and we've all heard now, for the umpteenth time I'm afraid, about gaps between rich and poor, and about ecological imbalances. My question has to do with the capacities to deal with those challenges. What about our human and natural capital? How can we make globalization, for instance, sustainable? I agree with many of the speakers: we do need new financial mechanisms; we do need new institutional arrangements; we do need new economic structures. I personally would like to add trade regimes, investment regimes, and capital flow regimes to the list of institutions mentioned. We also need changes in lifestyles and we need ethically driven codes of conduct for corporations. But I think first and foremost we need more—and particularly more widely spread—awareness, ideas, knowledge, know-how, will to manage and govern, and participation in development to make it more human and more sustainable.

However, there's a problem. Of course, in terms of primary and secondary education we are becoming aware of the need to raise the levels of these and to raise literacy. Yet UNESCO tells us that when it comes to higher education and professional training, over the next decade there will be a growing gap between the need for institutions that provide such training and the distribution of institutions able to meet that need. There will be—there already is—a growing inequality in access to professional and higher education. This is not only unfair to future generations in developing countries, it also increases the risk of increasing unsustainable use of resources on this planet, with high costs to humanity as well as to biodiversity. I'm very happy that Ruud Lubbers highlighted this particular point in your thinking, Your Holiness.

UNESCO asks for ethical considerations to be given more prominence in education. After all, the citizenship, leadership, and management capacities of tomorrow will come from the next generation. Higher education, or education in general, can contribute to better governance, to stronger civil society, and to more socially responsible enterprise. So I would like to add the role of education to the list of instruments to effect change that have been already identified by the panel.

Let me reduce it to one issue: I would like to call for an increase in the capacity for higher education training in developing countries, to be facilitated not only by those countries' governments, but also by public–private partnerships from the West and the international community. I would like to call for more explicit and extensive plans to invest in these capacities, by bilateral and multilateral donors, including the World Bank and regional banks, through three things: institutional capacity building, fellowship programs, and the facilitation of widely accessible and needs-based distance learning facilities.

Ruud Lubbers: First, I wish to highlight what you said in relation to the definition of ethics, Your Holiness, and about inclusive thinking, taking into account the whole and the long-term perspective. That's the first point: the necessity of inclusiveness. That's basic. If we can apply that principle to education in order to understand what education is about, then we've already won one battle.

The other battle, of course, is the application of this principle. Throughout history, every nation and people has understood that education is vital. What's new now is that we are living in a *global* world, a *global* economy. We have to learn to set aside the financial means to ensure education and training all around the globe. But it's not just a question of money, it's also a question of partnership, as the speaker indicated. So we have to learn two things: to see the future of education without getting lost in the detail of decision-making; and to view education at a global level. That's where information and communication technology offers both a new perspective and an opportunity. We are becoming "emailing societies" and in these emailing societies we are connected with other people. We need to provide for their essential education as well. There will be no peace in the future without global education.

Fred Matser: We all *think*, but we forget so much about our *feelings*, and about allowing ourselves to be inspired, to be creative with one another, through time and space, with our Creator. One little thing on the subject of competition. Tennis is believed to be a game where the understanding once was that you did something together. The players hit the ball back and forth with the shared intention to keep the ball in play. It reflected nature, the cycle of life, co-existence, joy, creativity, variation, and so forth. And what did *we* make of it? We turned it into a competitive game in which the person who is the first to break that cycle is rewarded with a point!

I think this illustrates something we all suffer from: a belief system that demands that we have to win at the expense of a loser. But the reality in time and space is that we all share the same room with all other people, animals, trees, nature. We are all together, we share time. The whole idea of competition, in my understanding, is a fallacy; it's just a misunderstanding in our mind. If we could slow down our mind, we might be able to understand.

One last point: perhaps it might be a good thing in business to go from "human having-ness" back to "human being-ness," into more "be-is-ness" in business.

The Dalai Lama: Wonderful! Beautiful! These are good examples of how we can broaden and change our perspective. From the new perspective we can develop solutions, for example by changing some rules and practices toward improved collaboration and long-term benefit.

2.7 Compassionate economics

Stanislav Menchikov: Nine years ago, I promised the Dalai Lama that I would write a book about the "compassionate economy" or "compassionate economics." I can report that I have written that book, called *New Economics*,[4] which really means compassionate economics.

Since I have discussed most of the propositions in this book with the Dalai Lama during previous meetings, I don't really have any questions for him today. I understand his position. In fact, in ten years I find he has become quite an economist! A much more developed economist than he was ten years ago; that's great progress to see. But I do just want to say one or two things, and make a suggestion.

---

4   For further information, see http://www.louwrienwijers.nl/compassionate economy.html.

New Economics is now being officially recommended by our Ministry of Education in Russia for economics courses in Russian universities. Now the kind of economics that I teach is quite different from the prevailing neoclassical model that is taught in universities. I taught at the Erasmus University for quite a number of years before I retired; I must say that the standard, neoclassical model is very different from compassionate economics and, I find, very different from real life as well. I think that an economy that is totally based on maximization of profit and utility really is in conflict with human nature. It does not reflect the prevailing patterns of human behavior. Human beings are partly motivated by egotism, of course they are, and even greed. But the neoclassical model that is taught in universities actually teaches students to be greedy; it inspires them to think of profit maximization as the main goal, to be "maximizers."

> An economy that is totally based on maximization of profit and utility really is in conflict with human nature

If you look around you will see that most people are not really maximizers, but what one might call "satisfy-ers": they want to satisfy their needs, and that means being in equilibrium with oneself, with other people, with society, with civilization, and with nature. It is a different kind of equilibrium from the one we talk about in economics, but it is a basic equilibrium. It's also reflected in the family: the relations within the family are mostly based on altruism and compassion. So most of our lifetime we are really altruists and are compassionate.

> Most people want to satisfy their needs, and that means being in equilibrium with oneself, with other people, with society, with civilization, and with nature.

In business too, efforts to achieve possible and not unattainable levels of growth guide many successful—and profitable—businesses, even in market economies. Any company knows that what it wants to do is preserve its own share of the market; but that's not "maximizing" actually, if you look into it deeply. I think we need to make considerable adjustments in the way we teach young people in our universities who are preparing to become socially responsible businesspeople, socially responsible citizens, and socially responsible consumers.

And with that I wish to thank the Dalai Lama for bringing me an insight into the way to think and teach economics.

Wessel Ganzevoort: When we discuss "changing the rules" we need to bring in the notion of leadership. In recent years, the Western world has reinvented the word "leadership." It's very often used as a label applied to what in the past we called "management." Of course, we all know leadership is about vision, about a future, about purpose—even purpose in life and the purpose of your company. But to my understanding, leadership is also very much about *alignment*. It's about aligning behavior, personality, ego, with the *soul*—I don't think we've used that word yet today—and the soul with the divine, God, the "beyond," etc. Leadership is very much about an authentic path of reconnecting, about knowing on a deeper level. And I believe—I think this is something Eckart Wintzen has already said—the world's organizations and nations can only change if individual leaders take a path of reflection, maybe even meditation and contemplation, and lead through reconnection and alignment.

However, to do this in a context and an environment that to a large extent is based on ego, power, status, competition, and money—a context I would say is brutal, aggressive, and selfish—isn't easy. I think many people that try to walk the path of alignment and reconnection take great risks. We dare not jump off the tiger that we have created together, because we fear that it will kill us. And we feel that our competitors should, at least, jump first.

So my question is: how do we restore trust—trust in society, trust within individual organizations, and, most of all, trust in the individual—so that the individual can grow and develop into an authentic and compassionate leader?

> How do we restore trust—trust in society, trust within individual organizations, and, most of all, trust in the individual—so that the individual can grow and develop into an authentic and compassionate leader?

Geoff Mulgan: In a sense, this question is similar to the one faced by governments: "How can we start cutting our arms spending when every other government is spending more?" and by businesspeople: "How can we act ethically if none of our competitors are doing so?" How can you create the confidence and trust for people to act differently?

The Dalai Lama: There's no simple answer. Basically, I believe that if you have a more compassionate attitude you find it very easy to communicate with your fellow human beings; likewise if you are truthful, honest, and open. These are, I think, the bases of trust. If one side tries to hide something and tries to deceive, how can trust develop? Very difficult. Openness and straightforwardness—they're the bases for trust.

2.8 How to create responsible markets?

Sander Tideman: The modern economy is driven by the powerful concept of free markets. Markets create much that is positive but also much that is negative: waste, pollution, and other so-called "externalities." The key principle is called the "invisible hand of the market" which is a central tenet in classical economics, currently embraced by mainstream economic thinkers, as part of the neoliberal economic ideology. However, in day-to-day reality it means that nobody seems to be in control. Today even governments don't have much influence; they too have put their faith in the principles of the free market. Even though it is undeniable that free-market economics is creating a lot of wealth for many people, there are also many people and even some nations who lose out. So how can we "manage" markets, so that they stop creating negative externalities? How can we develop a system that moves from blind faith in the invisible hand of the market, to adhering to principles that would ensure a more responsible outcome from the market? What can the teachings of Buddhism contribute to redirecting the economy so that all sentient beings benefit?

The Dalai Lama: That's also difficult to say: this is a complex topic. Some people have coined the expression "Buddhist economy"; I'm not sure what that means. Of course, for Buddhists, just as for practitioners of any other religion, contentment is an important practice. But as far as the global economy is concerned, I don't know. In any case, as I mentioned before, a serious matter is the gap between rich and poor at both global and national levels. The gap between rich and poor grows bigger and bigger.

A few years ago someone told me that the number of billionaires in America had reached 40 or so. Another friend in Chicago told me recently that the number of billionaires is now 400. Is this figure correct? Do any of you have any idea? Anyway, the number of billionaires in America is clearly increasing while many people remain poor; in some areas of the country access to even the basic necessities is inadequate. We can see the contrast between the very rich and the very poor among the inhabitants of just one big city like New York. This is very sad, and not only morally wrong, but also a source of practical problems. If one lacks even the basic necessities, it's very difficult to practice compassion and caring for one another.

So, we have to address this inequality, and in this respect I think socialist ideas are very relevant. I remember an occasion in India when members of a rich family came to seek a blessing from me. "Oh, I cannot bless you," I told them, "I've nothing to give." Of course, real blessing is an inner act and not a matter of giving any *thing*. I went on, "As a rich family you have used *capitalism* to reap your profits; why not use *socialism* to distribute these profits to improve the health and education of the poor?" That was my suggestion: to make money by exploiting capitalism and then spend it by applying socialist principles. I think that is proper. Actually, I was also told in America that in the past, rich people were very stingy, while nowadays more of them give donations to help with the education and so on of underprivileged sections of society.

> As a rich family you have used capitalism to reap your profits; why not use socialism to distribute these profits to improve the health and education of the poor?

At a global level, there is a big gap between the Northern industrialized nations and under-developed Southern nations. According to some experts, the Southerners' standards of living have to be raised. If their living standard is raised to the standard that Northerners already enjoy, it's questionable whether natural resources will be adequate to meet demands. Just imagine, if every person in India and China acquires a car—that's 2 billion cars! Problematic, isn't it? Living standards will rise in general, so sooner or later the Northerners' lifestyle will have to change in line with new imperatives. In the West, you have a deeply embedded expectation that in order for the economy to be successful, there has to be growth every year; sooner or later you will discover that growth has its limits. These are very serious matters.

> In the West, you have a deeply embedded expectation that in order for the economy to be successful, there has to be growth every year; sooner or later you will discover that growth has its limits.

Sander Tideman: I think here we need to explore another related aspect of modern market economics: the culture of consumption. Buddhist teachings say that greed is an afflictive emotion, that is, an emotion that is negative and disturbs our peace of mind and therefore should be discarded from our mind. In our Western market economy, however, consumption is

PHOTO: LAMBERT VAN DER AALSVOORT

Sander Tideman addressing the "Compassion or Competition" dialogue.

encouraged because, without it, our economy does not grow fast enough. Because of the encouragement of consumption—mainly through marketing, advertisements, and commercials—you can see that greed is also encouraged. People are told that to have desire for the newest fashion, the newest products, is natural. What would you say to modern economists who propagate the economy of consumption, and therefore encourage greed?

The Dalai Lama: We should seriously think of the environmental consequences of a consumerist society. Although advances in science and technology may be able to help us adapt to some degree to the consequences, in the long run we will have to face limited reserves of resources and the prospect of reaching a point where even science cannot rescue us. As a society, it is important to practice contentment, so that our greed and excessive consumption are not constantly overtaking us. At the level of the individual in particular, it is important to realize that, no matter how far we go in trying to gratify our greed and desire, we will not find total satisfaction. Rather, fulfillment is found by adopting the inner disciplines of self-restraint and a sense of modest needs.

2.9 What can you do as an individual?

Question from young business leader: Your Holiness, I'll try to be as "young and naïve" as I can. I need your help. I work in the business sector and I'm an idealist who wants to make a difference to the world in the ways that all the panelists mentioned. I live and breathe the business sector mentality and environment, and I perceive that there is a sea change needed of a magnitude that far outweighs the energies that I have to give. So my question is: could you please give me some directions that will help me to keep on trying to make a difference?

The Dalai Lama: I don't know. My own experience has been that no matter how numerous the difficulties and no matter how big the obstacles, if your belief or your ideal is truly reasonable and benefi-cial, then you must keep your determination and maintain a constant effort. I also think that if something is right and good for the larger com-munity, then whether that goal materializes within one's own lifetime or not doesn't matter. Even if it's not going to materialize within our own lifetime, we have to keep working. The next generation will follow and, with time, things can change.

> If something is right and good for the larger community, then whether that goal materializes within one's own lifetime or not doesn't matter. Even if it's not going to materialize within our own lifetime, we have to keep working. The next generation will follow and, with time, things can change.

Here, I think the main problem is certain ten-dencies, whether right or wrong, already deeply rooted in society or in the mind. My impression is that the whole structure of society is devel-oped in such a way that one individual, or a few individuals, cannot do much to change things. I notice some young university undergraduates, for example, have a very fresh mind, full of ideas, ide-als, and enthusiasm when I first meet them. Then, a few years later when we meet again, they have their own family and they've become part of the whole wheel, going round and round and round, complaining as they do so. But you've no choice but to go with that cycle! I think they simply begin to realize that something is *lacking*, something is *wrong*, and with that real-ization, I think one does some kind of groundwork. Particularly in the field of education, I think we need to make clear and emphasize to the younger generation the basic human values and the importance of a warm heart.

Gradually, through evolution, I think there is thus a possibility to transform, to change.

So, I'm not expecting that certain among my fundamental beliefs will materialize within my lifetime. But that doesn't matter. What's important is to clarify what is wrong and to make a start.

Geoff Mulgan: Perhaps I could ask other speakers to very briefly attempt to answer the same question: what *practical* step would you offer to our questioner?

Hazel Henderson: Two short points. First, we should ask our professors of economics in all universities to teach about the altruistic, unpaid half of the economy. I have been writing about it for 25 years and every woman in this room, as well as most men, knows all about it.

Second, it is my view that we can move competition into new areas. For example, in the U.S.A. the competition now is to see who can give the most money away. Will it be Bill Gates, or will it be Ted Turner? So let us encourage this kind of competition!

Jermyn Brooks: I certainly don't know the answer. But for individuals seeking the right path and the way to influence the world in the right way, let me just say from a business perspective, that I don't apologize for supporting the concept of enlightened self-interest in business activities. I think if it's in the enlightened, long-term, sustainable self-interest of business, then we will find that the activities that that business develops are of benefit to us all and not just to those people who are involved in those activities. So I would encourage the questioner to the extent that her future in business is to embrace those principles. And I'm actually quite optimistic.

One of the reasons I'm so optimistic is precisely because of this conference. Business would not have been invited to this kind of conference only a few years ago, so things are moving and businesspeople are now stepping up to the challenge of ethical behavior without embarrassment, much more so than they were only a few years ago. Just a small example in terms of the educational challenge which one of the speakers raised. We were recently asked to talk about sustainable development to the London Business School; this was not an official invitation, it was an initiative by the students. As a result of that initiative, the London Business School—as just one example of a business school—has taken on a full program which is fully part of

PHOTO: LAMBERT VAN DER AALSVOORT

The Dalai Lama meeting crowds in Amsterdam 1999.

the credit system for the MBA program, in which they will talk about ethical business, stakeholder value, and the issues we've been talking about today.

So I think we are creating change and I think the individual and all of us can make a difference if we believe in this direction.

Ruud Lubbers: My observation is basically that people are very scared of what's called "globalization." They are scared, they are concerned, and they are becoming very cynical. Hazel is right: sometimes you see new things, such as very rich people who gift money to charity and perhaps this is a blessing in disguise. There are new initiatives as well, such as Médecins Sans Frontières, as I already mentioned.

But now I want to conclude on my part by linking up with what His Holiness said when answering a question about trust; that is, that it's so important that people perceive that things are done in good faith, truthfully, and with dignity. Personally, I believe that politics, business, and civil society are all important, but what is even more important is that all three practice principles of transparency, accountability, and integrity. If leadership in the world—in the form of NGOs, companies, or politicians—can make it clear to the public that they are working according to standards of integrity, by working openly, transparently, and accountably, maybe the public at large will start to be less cynical and scared about what's happening today.

I realize these are the words of a former politician who was in the business of change—yes, politics is about change—but I defend the thesis. Having listened to His Holiness the Dalai Lama, I think that in order to practice compassion and relate to other people in terms of companionship, the basic requirements are a change in attitude and a change in practice—it's not just about nice theories.

I was very impressed when His Holiness explained that spirituality is not separate from things that are important for human beings and humanity. It's exactly in the interests of human beings to be spiritual and to see things from the perspective of the longer term. But to do this in a practical way—and once again, as a former politician, I'm biased—we have to capitalize on the energy and power of companies. We have to ask our civil societies to be accountable, to work transparently and with integrity. If we can manage to do that, we can bridge the gap with a spiritual and inclusive approach. And that, bearing in mind the question my Anglo-Saxon friend put to us, *is* practical.

Geoff Mulgan: I would like Your Holiness to have the last word. But just before bringing you in, I'd like to make three very brief comments on the discussion.

The first is that the founder of modern economics, Adam Smith, wrote *The Theory of Moral Sentiments*,[5] about compassion and sympathy, before *The Wealth of Nations*.[6] We need to reintegrate ethics and economics; they have become divorced unnaturally.

Second, businesses around the world are accumulating more power, not only I think because they want power, but often as a by-product of the search for markets and profits. With that power must come new responsibilities.

Third, we have a New Economy which is creating new inequalities and divisions as well as opportunities, which make it incumbent on governments to find new ways of including people, and incumbent on us as individuals to exercise compassion. Above all, it's vital that as the world becomes smaller we don't also develop smaller minds, shorter attention spans, and reduced senses of responsibility. What this evening has led me to think about, and

5   Smith, A. (2005). *The Theory of Moral Sentiments* (S.M. Soares, Ed.). São Paulo, Brazil: Metalibri. (Original work published 1790).
6   Smith, A. (2008). *The Wealth of Nations*. New York, NY: Oxford University Press. (Original work published 1776).

what I understand many of your teachings to be focused on, Your Holiness, is how we develop a bigger sense of our self, our place in the world, and our place in time.

I hope you will give us some final thoughts, Your Holiness, to take away with us this evening, coming out of the conversation.

The Dalai Lama: I've nothing to say … Except thank you! I really enjoyed it. Thank you very much.

Perhaps just one thing … I found that all of you who spoke simply expressed what you feel. I think that's very important. Sometimes in these meetings people make some sort of nice statement or speech, but you're left wondering: does that person really feel like that or not? That's not much help. In this forum we've been discussing some of the problems and challenges that we face, things that nice statements alone won't solve. We need a sense of commitment, and then we need to discuss matters seriously, sincerely. So, I found that kind of atmosphere here, and that makes me very happy. Thank you.

Part 2:
Designing an economy that
works for everyone

The author meeting the Queen of Bhutan at the first "Gross National Happiness" conference in 2004.

3
On the path of purpose

This chapter describes my work after the first public dialogue with His Holiness the Dalai Lama. At first I "lost" my purpose in corporate banking, but then I rediscovered it in sustainable investing, social entrepreneurship, and consulting. These new activities were inspired by a series of events. First, I returned to China as a social entrepreneur to help the Tibetan people in sustainable development. Second, after meeting the American businessman Anders Ferguson, we were able to continue the "Compassion or Competition" dialogue on a global scale through conferences in New York and elsewhere, and in the process a global network, Spirit in Business, was established. This network exposed me to the inspiring world of the modern science of the mind and social innovation. Third, the government of Bhutan, a secluded Buddhist Kingdom in the Himalayas, invited me to work with them in operationalizing their unique development model of gross national happiness. In the ensuing years (1999–2004), as I traveled the world in a new identity, I had a genuine sense of living my purpose, a phase characterized by exploration and discovery. But it was also a somewhat chaotic time, leading to new levels of confusion.

3.1 Times of change

Initially, it was not quite clear what the "Compassion or Competition" forum in Amsterdam had brought about. While all participants were inspired and encouraged by the exchange with the Dalai Lama, I still struggled with making sense of integrating the complex dynamics of globalized economics with values, ethics, and sustainability. It dawned on me that we had to conduct research on at least three levels: macroeconomic systems, business organizations, and personal leadership. Presently, these levels were all lumped together, and the debate about alternative approaches to business and economics became confusing and ineffective. The scope was broad and diverse, so I did not know where to start.

I also faced a number of personal challenges. I had decided to leave ABN AMRO Bank. The lack of focus on sustainability and my serious doubts about the general direction of the bank undermined my morale at work. To continue specializing in a narrow field of financial engineering merely to make money for distant shareholders did not appeal to me. In a strange twist of fate, the meeting with the Dalai Lama was also to my disadvantage. A senior leader in the bank regarded my public appearance with the Tibetan leader in Amsterdam as a poor political judgment, fearing that my meeting with someone who was regarded as a dissident by the Chinese regime would discredit both myself and the bank in the eyes of China. As a China expert representing the bank in China, I should have known better. Clearly my own values and convictions clashed with those of the bank. Then, one morning on a gloomy winter day, when I was getting stuck in a traffic jam traveling from my home to Amsterdam, I stopped my car and said to myself: "I can't do this any longer."

With a lifelong contract, good fringe benefits and three kids to take care of, the decision to leave the "golden cage" of corporate banking was not easy. After some serious discussions with my wife and a few months of deliberation, I decided to take the leap. By accepting a large salary cut, I was able to accept a position at Triodos Bank, a small innovative bank committed to the values of sustainability. With roots in anthroposophy, the bank was a pioneer in funding projects in organic farming, renewable energy, culture, and healthcare, with a focus on small enterprises and microfinance. My job was to help expand the bank's international presence, which suited my pioneering drive and interest in emerging markets. Even though the salary cut was painful, there was a surprising upside. It was a big relief to work in an environment aligned with my values. I experienced this on a visceral level: I slept better, I laughed more, and I was much happier company for my wife and kids. Only then did I realize the price that you pay when your work environment contradicts your values and beliefs.

Even though economic ideology assumes that we are driven by material desires, I experienced the opposite in my life. The Dalai Lama had talked about the importance of motivation: I had now discovered that making money was not enough for me—I wanted to make a positive contribution to society. Even though I did not know exactly how, I felt it was now time to "maximize" meaning. I decided to surrender to this inclination to make a difference, a phase which I describe as "**living your purpose.**" Life took on an entirely new quality. Everything became interesting. Everything became an exploration. In a way, purpose led me. A seasoned social entrepreneur said to me: "Ah, you have been infected by the MAD virus!" (MAD standing for "Making a Difference"). Indeed, this experience of attempting to live your purpose was rather confusing! Many new opportunities arose, but sometimes it was too much. Every week I encountered new people and openings; I had a real hard time saying no!

3.2 Maximizing meaning

I had negotiated with Triodos Bank so that I could continue my work for the Tibetans and Central Asians, which had not stopped since returning to Holland. In fact, the discussions I had with the Dalai Lama and his office on the development of Tibet had continued over the years. These culminated in an invitation to support the establishment of The Bridge Fund,[1] an initiative of the Tibetans to bring foreign aid into the Tibetan areas of China in order to bring about sustainable development on the plateau, with a focus on enterprise development and cultural empowerment. By then the international donor community had realized that Tibetans inside China needed help and the Tibetans responded by creating this fund to serve as a "bridge" between the West and Tibet.

My experience as a banker in China was useful in setting up a structure that, on the one hand, would be in line with Chinese law and policies, and on the other, allowed funding to reach target populations in remote locations of the Tibetan plateau. I felt very honored to be able to serve His

---

1   For further information, see http://www.bridgefund.org.

Holiness in this way, but I knew that it was not without difficulties and risks, given the continued political tension between the Chinese government and the Tibetans. We needed to make sure that from the start we would be formally disassociated from the Dalai Lama and work autonomously. But it was a challenge I loved to take on. My experience in China had given me both insight and inroads into the mysterious workings of Chinese bureaucracy and business. My work with Mongolia and other countries in Central Asia had given me a grounded perspective on societies which have a semi-nomadic past and are in the complex process of reconciling their cultural values with modern development and the rise of consumerism.

I was excited by my new life of a sustainable banker, consultant, and social entrepreneur. Even though by then I had seen much of the world, my new work opened up a completely different environment to me, beyond the lobbies of five-star hotels, eight-course banquets, and business-class airport lounges. I now found myself on long drives over the Tibetan plateau, Mongolian steppes, the Xinjiang desert, Indonesian rice paddies, and Thai forests to meet dedicated environmentalists, small entrepreneurs, social workers, and groups of indigenous women. I spent several months working on supporting microfinance institutions in Mongolia, Cambodia, and India. These field trips were interspersed with calls into the world's major capitals, often in the slipstream of the Dalai Lama, to raise funding from Western donor agencies for the Tibetans.

But I also felt a bit overwhelmed. What to focus on? What was most impactful? I loved working at the grass roots in developing countries, but there was also the dialogue on larger issues with the Dalai Lama that I had started. I strongly felt that the whole system of capitalism needed to be rethought. It was good to channel funding into efforts to preserve cultures and ecosystems, but in a way it seemed nothing but cleaning the deck of a supertanker that is fast approaching an iceberg. How could I contribute in a more strategic way?

3.3 If it has benefits, then do it

When I met the Dalai Lama again, I was keen to get further guidance on where to focus my attention. It was hard to get a meeting in his schedule, especially when he was traveling. I was given the last time slot on his visit to Oslo, which meant that I could see him for some 15 minutes in his hotel room before his departure to the airport. As usual, I felt a mixture of excitement and shyness before the meeting with His Holiness. I knew it was impossible to raise my issues in such a short meeting, but there was also joyful anticipation that, as had happened in all previous meetings with the Dalai Lama, some new perspective would be revealed. While his assistants were busy vacating the room, he asked me to sit on a chair in his suite and listened intently to what I had to say. It did not take long before he interrupted and said:

"You know, the most important thing is your motivation. If what you want to do has benefit, you should try to do it. No hesitation. But all this requires a long-term vision. Helping the Tibetans is especially beneficial because it also helps the Buddhist teachings to survive—the culture of Tibet is the repository of the Dharma. This is the real purpose behind why I escaped my country—the Chinese prevented me from helping the Tibetans. I could be of greater help by escaping to India. So, if you want to support my purpose, whatever small contribution you can make for this purpose is much appreciated. Now, aside from Tibet, if you really feel that you can make a difference by raising these larger questions about economics then you should do it. But don't expect immediate results. And do things with small steps, take time to reflect and think. Take the middle way. I will join you on meetings when I have time."

While his assistant was reminding His Holiness of the time, the Dalai Lama brought up a different issue, at least different to my mind. "Tell me, do you think that China joining the World Trade Organization will have positive or negative consequences?" I had not expected this question, so I replied quite incoherently that it had pros and cons, that it would perhaps be good for overall trade and investments in China, but that politically this may be seen as a setback for the international lobby for human rights and the Tibet issue, a kind of condoning Chinese policies by the international community. "So I don't really know," I concluded, shyly. The Dalai Lama laughed and then, after a pause, concluded: "So if it will bring benefits to the majority of Chinese people, then we should support it."

He took my hand and we walked to the elevator where his entourage was waiting, and we all squeezed into the small cabin. On the way down, His Holiness continued to hold my hand. When the doors opened he let my hand go and waved to a large crowd of Norwegian fans, stepped into his car and disappeared.

It was the combination of the open and focused mind of the Dalai Lama that impressed me most. His mind could move from the big picture issues to very concrete matters, and vice versa, without any stress or effort. It seemed that there was no distraction or anxiety in his mind, which served as a reservoir of inspiration for good. Even the two-minute ride in the elevator was used to warm my heart by holding my hand. When we left the elevator, his mind instantaneously opened to all well-wishers awaiting him. His battery to generate warmth and optimism around him was always fully loaded. It was a strong battery full of tangible power. There was nothing soft or weak about the Dalai Lama.

Lodi Gyari, the special envoy of the Dalai Lama in the West, who was entrusted with the negotiations with the Chinese, once told me how he had experienced the Dalai Lama's compassionate power. In 1989, thousands of Chinese students had gathered at Tiananmen Square in Beijing to stage a nonviolent protest against the lack of democracy in China. The Chinese leadership responded with a violent crackdown and hundreds of innocent students were killed. His Holiness was appalled: he wanted to issue a statement in support of the students and rejecting the unnecessary violence. As this was at the time that Beijing was finally considering, after many years of negotiations, to allow the Tibetans more autonomy, Lodi Gyari hinted that the Dalai Lama should reconsider the issuance of the statement, as he feared it would jeopardize the reconciliation efforts with the Chinese. The Dalai Lama had looked at him fiercely, saying: "The reason that we want more freedom in Tibet is because we want to express our values of democracy, openness, nonviolence, and equality, as our fundamental human right. When the Chinese students are calling for exactly those values, I have no choice but to support that, in spite of temporary setbacks from the side of the Chinese regime." This was a powerful show of the Dalai Lama's mind-set to unreservedly support whatever is of benefit, regardless of consequences for selfish concern. For me this was the hallmark of a great leader.

These words, "If it has benefits, then do it—don't expect immediate results—take the middle way," became my mantra for the next few years. Clearly I was trying to do too many things at one time, but now I knew that as long as it had benefits I should not give up. I trusted that with time it would be clear what I should focus on to achieve

> If it has benefits, then do it—don't expect immediate results—take the middle way.

optimum impact. I started to practice the "middle way" for myself, by taking some time off and doing some longer Buddhist retreats. It was slowly beginning to dawn on me that, if I wanted to change the world, I should begin by changing myself.

3.4 Spirit in Business

In the fall of 2000, I received a mail from the American businessman Anders Ferguson. I will never forget his determined look when he said in our first meeting in New York: "The discussion that you started in Amsterdam is exactly what we need to do in order to create a sustainable world, so I want to help you to hold this type of dialogue on a global level." Anders had run companies and investment funds, but then he had made a journey to inner China and taken up the practice of meditation, which shifted his perspective on business. He had seen how globalization was uprooting ancient cultures and ecosystems. Like me, his mission became to help business to become a force for positive rather than negative change. He said he would devote the rest of his life to that transition. I could not resist Anders' call: I joined him and his team to create the Spirit in Business conference in New York, in April 2002. I invited my friend Marcello Palazzi who had helped me with the "Compassion and Competition" forum. Together, we created the first global effort to align business and economics with values and ethics.

One day, after returning from New York where we had committed the funds to launch the event, I turned on the television to see the shocking image of planes crashing into the Twin Towers. Just two days earlier I had slept in a hotel adjacent to the World Trade Center, which—including its hundreds of occupants—had vanished in smoke. For a few days, we were so distressed that we felt that the plan to hold a major business conference in Manhattan was unviable. But the shock of 9/11 did not deter Anders and his team.

As firefighters and rescue workers slept and drank coffee in Wall Street's favorite spiritual home, Trinity Church, a new spirit emerged as people tried to make sense of the senselessness. The idea of Spirit in Business (SiB) now enabled some businesspeople to reveal their own spiritual struggles. As New York City began to heal, the sense of urgency to find meaning in the insanity of 9/11 also convinced our team that we could host a global conference in this broken city in six months. They had developed the conviction that our conference was the very thing that was needed in New York. In fact, in the post-9/11 era, the choice of New York as a venue proved to be significant. It was the meeting point of two extreme opposites of globalization: global capitalism and global terrorism.

We were aided by cracks in the system that erupted at that historic time: the collapse and frauds of WorldCom, Tyco, Enron, and Parmalat precipitated loss of confidence in big business. Such misconduct brought us the

Sarbanes–Oxley Act of 2002, and turned disgraced CEOs—Ken Lay, Dennis Kozlowski, and Bernie Ebbers—into household names. WorldCom's July 2002 bankruptcy was the world's largest at the time (US$103.9 billion).[2] That is when public confidence in business really started to ebb. This coincided with the bursting of the dot.com bubble, which destroyed any illusion that the highly acclaimed New Economy was going to provide solutions to the world's problems; growth in the real economy started to stagnate. Simultaneously, continued globalization, new internet-based technologies, and financial innovation created new markets and growth. It was a time of both fear and hope.

Some world-recognized thought leaders decided to support us: Peter Senge, the founder of Society of Organizational Learning,[3] David Cooperrider, father of Appreciative Inquiry,[4] and Daniel Goleman, who had published the ground-breaking book *Emotional Intelligence* just five years previously.[5] Spiritual leaders came from a wide range of traditions and parts of the world. I invited the Dalai Lama to join, but unfortunately he had to cancel his participation due to a medical issue. He sent his envoy, Lodi Gyari, instead. Scientists and thought leaders connected the dots between the theory, science, and practice of human transformation and systems change.

One of the most popular thought leaders was Dr. Jon Kabat-Zinn, teaching us that we can lower our stress, improve our health and transform our leadership through meditation.[6] Through his work at UMass Medical Center, he gathered leading scientists and senior Buddhist teachers to explore the connections between science and contemplative practice. This work began mapping scientific terrain that was not well understood in the West, but fundamental to human understanding of how our minds really work.

Interestingly, some of the most powerful co-creators of the conference were top management consultants working with some of the most

---

2   Fortune (2009). The 10 largest U.S. bankruptcies. *Fortune.* Retrieved from http://archive.fortune.com/galleries/2009/fortune/0905/gallery.largest_bankruptcies.fortune/.

3   Senge, P.M. (1990). *The Fifth Discipline: The Art and Practice of the Learning Organization.* New York, NY: Doubleday Currency.

4   The Appreciate Inquiry concept was introduced in the book David Cooperrider co-authored with Suresh Srivastva: Suresh, S., & Cooperrider, D.L. (1998). *Organizational Wisdom and Executive Courage.* Lanham, MD: Lexington Press.

5   Goleman, D. (1995). *Emotional Intelligence: Why It Can Matter More Than IQ for Character, Health and Lifelong Achievement.* New York, NY: Bantam Books.

6   Kabat-Zinn, J. (1990). *Full Catastrophe Living: Using the Wisdom of Your Body and Mind to Face Stress, Pain, and Illness.* New York, NY: Bantam Books.

progressive executives. They wanted to see the expansion of these ideas in the broader business community. Remarkably, they brought very senior leaders from several of the world's largest companies. Not only did business leaders take the risk of showing up and speaking, they often brought their companies along as sponsors: American Express, Forbes, Verizon, Keurig/ Green Mountain Coffee, and Unilever.[7]

So with the weather bright and the temperatures mild, 600 business and civil society leaders convened in April 2002 for the conference. Of those participants, 150 carried foreign passports. It was, frankly, a bit of a miracle.

The SiB conference became the seed of initiatives such as the Spirit in Business network, the Fowler Center for Business as an Agent of World Benefit at Case Western Reserve University in Cleveland,[8] the Spirit at Work Award,[9] and a whole range of follow-up conferences and seminars: São Paolo 2003, San Francisco 2003, Bangkok 2003, Zurich 2004, and Bhutan 2004. It seemed that a movement was born.

The SiB conference made clear to me that my quest was no longer one of a puzzled Buddhist banker, but one that represented a global mission for positive change by integrating the creative force of business with genuine sustainable development objectives, one which resonated with many people.

Another key insight was the fact that we could increasingly substantiate the key ideas of spiritual leaders and their wisdom traditions through new concepts from science. This was absolutely essential because, as we learned, you could engage in intimate discussion with nearly any leader of a Fortune 500 company about how they integrate values into their work life, provided you used science as a framework and avoided religious connotations. For most people business is a secular enterprise; people's faith is personal. Since science is the framework of business and technology, it became relatively easy to use language of the new "mind science" to connect with the business leader's mind and heart.

---

7 For more information, see the afterword of this book by David Cooperrider.

8 Today the Fowler Center for Business as an Agent of World Benefit features prominently in the Weatherhead School of Management at Case Western Reserve University in Cleveland.

9 The Spirit at Work Award, founded by Judi Neal, later turned into Edgewalkers (http://edgewalkers.org).

3.5 Revolution in the science of the mind

The SiB network exposed us to a wide range of scholars and scientists who operated in the field of connecting business and spirituality. A primary source was the Mind and Life Institute, which emerged in 1987 from a meeting of three visionaries: the Dalai Lama, the lawyer and entrepreneur Adam Engle, and the neuroscientist Francisco Varela.[10]

While the trio understood that science had become the dominant framework for investigating the nature of reality—and this modern source of knowledge that could help improve the lives of humans and the planet—the three regarded this approach as incomplete. Whereas the natural sciences rely on empiricism, technology, "objective" observation, and analysis, the founders of Mind and Life perceived the limits of any understanding that excludes the subjective dimension of human experience, meaning, and culture. They were convinced that well-refined contemplative practices and introspective methods could and should be used as equivalent instruments of investigation—instruments that would help create a more complete picture of the phenomena of mind and of life as well as a more humane science.

Through these dialogues, the Mind and Life Institute had catalyzed the birth of a new scientific field of inquiry: "contemplative science."[11] The findings in this rapidly evolving field, in conjunction with new insights in physics, psychology, biology, and economics, placed everything I had learned about economics at school and in business in a different light. It relegated the Homo economicus to the distant past (where it came from) and gave rise to a much more hopeful perspective for humanity.

In the field of **physics**, breakthroughs of quantum and astrophysics gave us access to a hidden world that was far more subtle and unpredictable than the world of Newtonian physics. One of the scientists who attended SiB was the noted physicist Hans-Peter Dürr.[12] In lectures that were both revealing and entertaining, he explained the implications of this work. At the most fundamental level, quantum physics postulated the string theory, which implies that everything—visible and invisible, material and mental, close and distant—is connected and entangled, in some sort of mysterious manner, beyond linear space and time. This opened the door to the study of consciousness, as it has shown to be relevant in the measurement process.

10   See https://www.mindandlife.org and http://www.mindandlife-europe.org.
11   See, for example, https://www.mindandlife.org/product-category/books.
12   For information about Hans-Peter Dürr, see https://en.wikipedia.org/wiki/Hans-Peter_Dürr.

Observations from different mental frames have yielded different measurement results in the quantum domain. In other words, it turns out that the mind impacts matter—mind over matter. In order to better understand this new reality, this field of science has gradually (though reluctantly) welcomed the views from spiritual leaders such as the Dalai Lama.

In **psychology**, a new field of inquiry arose called positive psychology, which began in 1998. In the first sentence of his book *Authentic Happiness*, Dr. Martin Seligman claimed: "for the last half century psychology has been consumed with a single topic only—mental illness."[13] For decades, the orientation of psychologists had been to see how they could get people to move from minus to zero. Instead, Seligman and his colleagues focused on understanding how humans can develop positive qualities that enhance their well-being and happiness. The question became how people could go from zero to plus. For some strange reason, psychology had ignored the positive potential of people. To describe the effects of cultivating one's positive mental potential, Seligman coined the term "the meaningful life." Such meaningful life comprises a number of elements expressed in the acronym PERMA: Positive Emotions, Engagement, Relationships, Meaning/Purpose, and Accomplishments.

At the crossroad between psychology and biology, the field of **neuroscience** gained much influence; thanks to new technology such as magnetic resonance imaging, insights could be based on data gathered directly from the human brain. By identifying neural correlates of consciousness (a specific system in the brain where activity correlates directly with states of conscious experience) for emotions and moods, neuroscientists can use methods such as brain scans to tell us more about the different ways of being happy. Richard Davidson, a leading neuroscientist who worked closely with the Dalai Lama, has conducted research to determine which parts of the brain are involved in positive emotions. He found that the left prefrontal cortex is activated when we are happy, and is also associated with a greater ability to recover from negative emotions and an enhanced ability to suppress negative emotions. Interestingly, Davidson found that people can train themselves to increase activation in this area of their brains.[14] It is

13   Seligman, M.E.P. (2002). *Authentic Happiness: Using the New Positive Psychology to Realize Your Potential for Lasting Fulfillment*. New York, NY: Free Press.
14   Lutz, A., Dunne, J., & Davidson, R. (2007). Meditation and the neuroscience of consciousness. In P. Zelazo, M. Moscovitch, & E. Thompson (Eds.), *Cambridge Handbook of Consciousness* (pp. 499-554). Cambridge, UK: Cambridge University Press.

thought that our brains can change throughout our lives as a result of our experiences, a phenomenon known as neuroplasticity.

Since then, numerous clinical and organizational studies have shown that positive emotions such as compassion, care, love, and trust create profound biochemical and neurological changes in our bodies that are inherently good for us, physiologically and emotionally. Through practices such as mindfulness, the brain can make entirely new neural pathways. Lower blood pressure, increased immunity, enhanced cognitive function, and improved communication skills are all examples of the types of benefit from these practices for the individuals involved. Not only do these mind-training practices enhance people's sense of their own well-being, but they also help people to become more caring and compassionate in their lives, which ultimately results in improved societal outcomes and more equitable responses to societal challenges.

There are profound implications for business and economics from these findings. In business we often assume that motives of power, greed, and selfishness—in short, the neo-Darwinian model of survival of the fittest—are the most basic elements of human nature. However, this perspective is now contradicted at the level of our existence as biological beings, where research has shown that we are at least equally driven by motives of care and altruism.[15] The fact that our mix of drivers and characteristics includes those that can be considered the opposite of the *Homo economicus*, namely compassion and altruism, is a revolutionary insight.

3.6 Motivational systems

Tania Singer of the Max Planck Institute in Germany, another neuroscientist and dialogue partner of the Dalai Lama, has done detailed research on human motivational systems, explicitly motivated by an aim to develop a more caring form of economics.[16]

15 Fredrickson, B.L. (2003). The value of positive emotions. *American Scientist*, 91, 330-335.

16 Singer, T., & Ricard, M. (2015). *Caring Economics: Conversations on Altruism and Compassion Between Scientists, Economists, and the Dalai Lama.* New York, NY: Picador; Singer, T. (2008). Understanding others: brain mechanisms of theory of mind and empathy. In P.W. Glimcher, C.F. Camerer, E. Fehr, & R.A. Poldrack (Eds.), *Neuroeconomics: Decision Making and the Brain* (pp. 233-250). Amsterdam, Netherlands: Elsevier.

Roughly speaking, there are three motivational systems in the human brain. One is a wanting and seeking system, or an **incentive- and reward-focused system**. This system is associated with wanting, achieving, consuming; with drive and excitement. Some of this can be considered healthy, some less so. This system evokes feelings of euphoria and wanting more.

Another system which is also vital for survival is what is called the fear system, or a **threat-focused system**. This is a system that is activated very quickly in our brain when we perceive danger. If you are afraid of snakes, for example, this system will be activated when you see a snake—you will scream and run away. The fear system is implicated in producing feelings of anger, anxiety, disgust, and panic; it can trigger a stress response in the body, which is associated with an increase in cortisol levels. Chronic fear can thus make us ill, but normally this system is adaptive, priming us with feelings of self-protection and ensuring we seek safety when we are in danger.

The third system—and in economics we tend to forget about this system—is the caring system, or the **affiliation-focused system**. Every animal has it. It is essential for mother–child bonding, for connectedness, love, and contentment.

Both the reward system and the caring system are associated with positive feelings, but the former is associated with high arousal and the latter with relaxation. What activates the caring system? In primates, grooming has been found to trigger this system. It is also associated with the release of a hormone and neuropeptide called oxytocin. Studies show that massage and intimacy activates oxytocin, which is why when you get a massage or a loving embrace, you feel calm and soothed.

The question is, of course, which of these motivational systems is the strongest and most fundamental? Traditional economists would argue that the most primal drive must be the reward-focused system, as it most closely matches the axiom of selfish and rational choice in free-market economics, and particularly suits the mantra of the modern consumerist society. However, the science points in a different direction.

This is evident from research in another promising field of study, **behavioral economics**. Experiments in this field are based on the actual behavior of people, not on theory or mathematical models, which we have made synonymous with modern economics. When combined with neuroscience, the resulting insights into economic behavior can uproot long-held axioms of economic theory. The theoretical and philosophical reasoning of the latter cannot stand up to the robust findings of neuro- and behavioral sciences.

For example, research by the microeconomist Ernst Fehr has challenged existing economic models based on selfish preferences, showing that people take fairness into account during economic exchanges. Ernst Fehr designed the "social dilemma" experiment, sometimes called the "trust game" (Fig. 3.1), which he presented at a conference with the Dalai Lama.[17] The experiment goes as follows: two strangers, say John and Sue, are paired anonymously. Each receives an endowment of $10. John can transfer between $1 and $10 to Sue. The experimenter then doubles that amount. So, if John sends $1 to Sue in this experiment, Sue will receive $2, while John retains $9. Now Sue can transfer money back, also between $1 and $10 and the experimenter again doubles this.

It is important to note that there is no incentive for John to give any amount to Sue. There is no guarantee that Sue will reciprocate John's generosity of $1, which causes him a loss of $1. Since John and Sue are strangers and don't meet after the experiment, there is no future potential gain from being nice to each other. In essence, this experiment represents a typical economic exchange that we witness every day: John has something that Sue values, and Sue has something that John values. If they exchange these, both are better off. The question is: can John trust Sue's motivation of reciprocating John's offer?

This game has been played by thousands of people in multiple settings and cultures. The results show that the first person allocates money to the stranger, thus reducing the amount that they have, while recipients tend to reward the trust that they have received by increasing the sum that they return to the first person.[18] In our example it would mean that, at average, Sue sends $2 back to John, which doubled by the experimenter becomes $4. As a result, by sending $1 to Sue and trusting her fairness motives, John gains $4.

What this experiment shows is that the average person is guided by an innate sense of fairness and care. This extends even to strangers whom they will never meet again, so their behavior cannot be based on a calculated motivation for selfish gains. Moreover, it shows that people have trust in each other's altruism—they know that others, like themselves, have altruistic motives. Of course, this does not preclude the possibility of reward- and threat-focused systems manifesting, perhaps overriding altruistic impulses.

17 Singer, T., & Ricard, M. (2015). *Caring Economics: Conversations on Altruism and Compassion Between Scientists, Economists, and the Dalai Lama*. New York, NY: Picador.

18 Henrich, J., Boyd, R., Bowles, S., Camerer, C., Fehr, E., & Gintis, H. (2004). *Foundations of Human Sociality: Economic Experiments and Ethnographic Evidence from Fifteen Small-Scale Societies*. Oxford, UK: Oxford University Press.

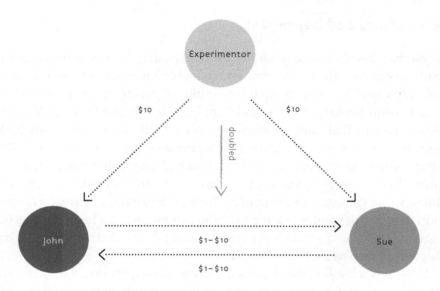

FIGURE 3.1   The trust game

Nonetheless it shows that, under neutral circumstances, the sense of caring and sharing is quite natural for us.

The outcome of this experiment is another blow to the *Homo economicus* model of economic behavior: if individuals were only concerned with their own economic wellbeing, they would allocate the entire good to themselves and give nothing to the stranger. In other words, altruism and care are not "externalities" to economics, as the classical economics theory suggests, but are essential (yet rather overlooked) economic drivers. In fact, there are plenty of real-life examples underscoring how economics cannot function without a degree of altruism and care. Ernst Fehr explains how altruism is relevant to economics:

> Altruism provides social insurance—[by definition,] altruists help when help is needed. In the absence of the welfare state, only altruism is left. In fact, you can argue that the welfare state itself is partly the result of altruistic efforts. Altruism also increases the volume of mutually beneficial economic exchanges. Why? Because we are more willing to keep obligations if there are people in society who behave altruistically, or who punish altruistically. Altruism helps to enforce the cooperative norms that are the basis of human culture.[19]

19   Singer, T., & Ricard, M. (2015). *Caring Economics: Conversations on Altruism and Compassion Between Scientists, Economists, and the Dalai Lama.* New York, NY: Picador.

3.7 Money and happiness

Economics revolves around money. Without money there is no business and without business there is no money. Where are happiness and well-being in this equation? Modern research has found surprising results here too. The relationship between money and happiness is not at all straightforward. While it is true that money makes a significant difference to the poor, it has a greatly diminished effect once one reaches a certain level of income. One study found money ceased to aid levels of happiness after a person makes over US$75,000 a year, and people overestimate the influence of wealth by 100%.[20] The professor of economics, Richard Easterlin, noted job satisfaction does not depend on salary.[21] Having extra money for luxuries does not increase happiness as much as enjoying one's job or social network; this has become known as the "Easterlin paradox."

If there is a relationship between money and happiness, it is in fact the opposite of what people think. Money is no guarantee of happiness, but happiness can lead to money, in the sense that it leads to success. The researcher Sonja Lyubomirsky reports: "Study after study shows that happiness precedes important outcomes and indicators of thriving, including fulfilling and productive work."[22]

It is quite astonishing how much emphasis is put on the generation of money in economics and business, while it has so little long-term effect on our well-being. What then generates or secures our well-being?

According to neuroscience, there is a certain section of the brain's frontal cortex called the rostral anterior cingulated cortex, which boosts the flow of positive and happy emotions. This section is more active in optimists when they imagine a positive future. In other words, when we use this cortex by imagining our desired futures, we are actually improving the flow of "happy emotions." This process also leads to actually making these imagined positive changes to our future.

20   Aknin, L., Norton, M., & Dunn, E. (2009). From wealth to well-being? Money matters, but less than people think. *The Journal of Positive Psychology*, 4(6), 523-527.

21   Easterlin, R. (2008). Income and happiness: towards a unified theory. *The Economic Journal*, 11(473), 465-484.

22   Lyubomirsky, S., King, L., & Diener, E. (2005). The benefits of frequent positive affect: does happiness lead to success? *Psychological Bulletin of American Psychological Association*, 131(6), 803-855.

These discoveries have a wide range of exciting applications. Here are a few for people working in business:

- You want to be an optimist in your work and you want optimists to work for you. Optimists are capable of working longer hours and tend to earn more. They even save more. We can train in optimism. When unfortunate or negative events occur in our lives, our brains have a neural mechanism that generates optimism.

- You want to foster positive emotions and kindness in the workplace. Demonstration of affection is useful, leading to better health and greater success. Also, happy, positive, and kind people are more successful in their work.

- The part of the brain that boosts happiness can be trained by using all sorts of meditation exercise, such as mindfulness, guided imagery, self-hypnosis, deep relaxation, outdoor activity, and listening to certain types of music. This contributes to improved productivity levels.

- Productivity itself helps to eliminate feelings of being unfulfilled, especially when we are truly enjoying the work we do. Therefore, employers should help people to do work that they like.

In summary, the findings of these new mind sciences have given rise to a new paradigm in economic thinking. Recruited by marketing companies, some scientists in this field developed the field of **neuro-marketing**, which was then used to advise businesses to design sales and advertisement strategies aimed at hitting the right motivational system, so that the consumer's brain would say "buy!"

I was more interested in another application of neuroscience. How could this field inform business leaders and regulators to enhance the societal purpose of business? How could we develop economic systems that would activate the innate positive and altruistic parts of the brain, which could facilitate collaboration and care? I believed that this was even more important than boosting sales: it would boost happy, compassionate, and wise people.

3.8 Gross national happiness

In the summer of 2003, I received a call from Karma Ura, Director of the Center of Bhutan Studies in Bhutan, a remote kingdom in the Himalayas with a Buddhist culture similar to Tibet. We had met at a conference some years previously where we had both been speakers: he spoke on gross national happiness (GNH), which was Bhutan's development philosophy, and I lectured on Buddhist economics, based on what I had learned from the Dalai Lama and the new sciences. At the conference, when Karma Ura received pushback from a skeptical audience of Western economists, I had come to his aid by arguing that GNH did not contradict GNP—the conventional economic development indicator—but rather represented a higher order of indicators that included and even transcended GNP. I added that there were many progressive economists who were rethinking GDP and were open to supporting GNH in its development; I was referring to the emergent movement we had started with Spirit in Business. Karma Ura had remembered my remarks and now he needed my help.

Later that year, I flew to Bhutan. My first impression of the country was that it was like a "little Tibet," as it shared many cultural similarities with Tibet, yet a closer look revealed a marked difference. While Tibet had been taken over by China and subjected to modern infrastructural development, Bhutan had been able to preserve its colorful and authentic culture and environmental integrity. It was really refreshing to be in a country without the frenzy of highways, traffic jams, shopping malls, and pollution. People seemed happy and content, greeting each other as we passed on the street.

The king of Bhutan first expressed the term "gross national happiness" in 1972. It is rooted in the Buddhist notion that the ultimate purpose in life is inner happiness. Being a Buddhist country, Bhutan's king felt a responsibility to define development in terms of the happiness of its people, rather than in terms of an abstract economic measurement such as GNP. Karma Ura explained: "The ideology of GNH connects Bhutan's development goals with the pursuit of happiness. This means that the ideology reflects Bhutan's vision on the purpose of human life, a vision that puts the individual's self-cultivation at the center of the nation's developmental goals, a primary priority for Bhutanese society as a whole as well as for the individual concerned."

Over the last three decades, as the Bhutanese slowly evolved GNH as a guiding principle, the country expanded both its network of roads and its area under forest cover. Health and education are free, and available to all Bhutanese. In order to protect its cultural strengths, Bhutan delayed introducing television and the internet until 1998. In 2004, it restricted the inflow of tourists to about 7,000 annually to avoid crowding its tiny towns and also to prevent too many feet from trampling over its sacred places. Bhutan believes that its indigenous culture is generally self-sufficient and has little to gain from conventional Western developments.

However, it was becoming increasingly evident that the small nation with 700,000 people could not ignore modern-day global economic realities, which have increasingly powerful cross-border and cross-cultural impacts. Free exchange of information, driven by the world's mass media and advanced communication technology, has continued to erode traditional borders. Thus, Bhutan has had no choice but to take up the challenge of the global economy, to help shape and steer these economic realities toward improvements in quality of life, rather than only in terms of wealth acquisition. Of course, the picture is not perfect. While people are not going hungry, about 25% of the population is reported to be undernourished.

The Bhutanese government decided that GNH should not remain merely a slogan, but that it should become a guiding force for day-to-day economic and political decision-making for Bhutan. In other words, said Karma Ura, GNH should be operationalized, and Bhutan needed Western experts to assist with this. An important condition was to find those experts who had an appreciation for Bhutan's background and philosophy. They were all to be brought together in a major conference in Bhutan.

Thus, my brief was clear. I first approached Hazel Henderson, a leading thinker on economics and development, who had been a keynote speaker at the 1999 conference with the Dalai Lama and a prominent voice in the SiB network. She wrote the following words to endorse the GNH conference:

> Now that globalization of markets has followed the erroneous dictates of narrow and often faulty economic paradigms targeted toward per capita averaged GDP growth, we are confronted with their growing "externalities." These include costs in wider poverty gaps and social exclusion, but also in continued erosion of non-money-based local livelihoods and cultures, as well as the extinction of other species and ecosystem disruption. Bhutan has taken visionary leadership by creating the GNH philosophy, which can become a much needed alternative measure of sustainable development that includes rather than excludes the very values that make life on this planet flourish.

As Buddhist leader, the Dalai Lama is much respected in Bhutan. Due to pressure from China he was not allowed to visit the country, but in his letter of support for GNH he applauded the Bhutanese for taking this initiative:

> On a national and global level, we need an economic system that enables the pursuit of true happiness. The purpose of economic development should be to contribute to rather than obstruct this goal. There are many serious drawbacks to the world's current economic

system, as evidenced by environmental degradation and the increasing gap between rich and poor ... For all the innovation and creativity in our economic activity, we have not succeeded in securing the essentials for existence for all human beings everywhere. We will be able to resolve the disparities we witness today and achieve lasting peace only if we implement compassion. We need to find ways of bringing compassion to bear in our economic development.

With so many endorsements, the Dutch government was prepared to sponsor the meeting, which turned out to be a true watershed. In February 2004, over 82 scholars and experts from 20 countries gathered in Bhutan's capital, Thimphu.[23] Coming barely two years after the ground-breaking SiB conference, I felt overjoyed by the opportunities that this presented: I was witnessing the same paradigm shift emerging in both private- and public-sector thinking. The international gathering in Bhutan came to a consensus that, while Bhutan's GNH endeavor is unique, the concept itself need not be restricted to either Buddhist societies or small homogenous countries. This marked the beginning of an international GNH movement.[24]

The government of Bhutan defined GNH by the following four pillars:

1. Good governance

2. Sustainable socioeconomic development

3. Preservation and promotion of culture

4. Environmental conservation

These have since been translated into the GNH Index, which comprises data gathered from nationwide surveys, of which two have been conducted to date. The State Planning Commission was renamed the Gross National Happiness Commission, and was charged with reviewing policy decisions and the allocation of resources in accordance with the GNH philosophy. In order to ensure continuity of that philosophy and to raise local and international awareness, the GNH Centre was set up as an independent NGO. GNH now encompasses the nation's socioeconomic development framework, a policy-screening tool, an index, and an educational awareness-raising process.

23   The proceedings of this conference were published in Ura, K., & Galay, K. (Eds.) (2004). *Gross National Happiness and Development.* Thimphu, Bhutan: Center for Bhutan Studies. Retrieved from http://www.bhutanstudies.org.bt/category/conference-proceedings/.

24   Follow-up conferences have since been held in Canada, Thailand, and Brazil: see http://www.grossnationalhappiness.org.

Did these measures work? According to a global study on subjective well-being conducted in 2007, Bhutan ranked eighth out of 178 countries.[25] In fact, according to this study Bhutan is the only country in the top 20 "happiest" countries that has a very low GDP.

Since avenues have been opened for GNH to spread from Bhutan, the idea has gained immense international popularity as an alternative development philosophy. In 2010, Ben Bernanke, chairman of the Federal Reserve, said in a public speech that the next level in economics is to create new measurements models capturing happiness as the purpose of economics, and that we should learn from Bhutan's GNH Index.[26] In 2012, at the initiative of Bhutan, the General Assembly of the UN even made the conscious pursuit of happiness a fundamental human goal in the resolution "Happiness: towards a holistic approach to development."[27] The UN Secretary General Ban Ki-moon stated:

> GNP has long been the yardstick by which economies and politicians have been measured. Yet it fails to take into account the social and environmental costs of so-called progress. We need a new economic paradigm that recognizes the parity between the three pillars of sustainable development. Social, economic and environmental well-being are indivisible.[28]

3.9 Designing an economy that works for everyone

Bhutan forms a hopeful reminder that rethinking the purpose behind economics is worthwhile. It serves as an example of how economic models can be built on a different, more wholesome paradigm and an encouragement

25 White, A.G. (2007). A global projection of subjective well-being: a challenge to positive psychology? *Psychtalk*, 56, 17-20.

26 Bernanke, B. (2010). The economics of happiness. University of South Carolina commencement ceremony, Columbia, SC. May 8, 2010. Retrieved from http://www.federalreserve.gov/newsevents/speech/bernanke20120806a.htm.

27 United Nations (2012, July 12). Resolution 66/281 on Happiness. New York, NY: UN General Assembly.

28 Ki-moon, B. (2012, April 2). Secretary-General, in message to meeting on "Happiness and Well-being" calls for "Rio+20" outcome that measures more than Gross National Income. UN Press Release. Retrieved from http://www.un.org/press/en/2012/sgsm14204.doc.htm.

for anyone interested in leadership for sustainable development.[29] Within the SiB network, we encountered similarly hopeful examples in the private sector, among small and big companies, all across the world.

However, while I immersed myself in all these positive initiatives, I could not close my eyes to what was happening in the rest of the world. Evidently, there were strong forces at play, especially at the level of the globalized financial markets. Global capitalism had become inseparably linked to the ideology of neoliberalism, which assumes that a mix of free markets and free capital flows will solve all our world's problems. My study with leading scientists and my experience in Bhutan taught me that this is a flawed ideology, too simplistic and woefully inadequate for solving future challenges. I felt that we needed to upgrade our economic and business models; we needed capitalism, version 2.0. This was humanity's next big challenge. My SiB partner Anders Ferguson and I felt compelled to play our part. So we decided to reengage the Dalai Lama in this conversation, together with a number of leaders in business who we believed were up to the task.

---

29   For anyone interested in applying these ideas, see Tideman, S.G. (2016). Gross national happiness: lessons for sustainability leadership. *South Asia Journal for Global Business Research*, 5(2), 190-213.

4
Second dialogue: Designing an Economy that Works for Everyone (Irvine, California, 2004)

Following the format of the first dialogue with the Dalai Lama, the 2004 dialogue focused on the question of how to create a more inclusive economy, touching on the themes of fair and responsible trade, the role of technology, employee empowerment, women in business, leadership, organizational change, and management education.

The Dalai Lama acknowledged again the important role of business in creating prosperity and societal well-being, but stressed the need for business to serve the social and development needs of disadvantaged people, as well as to protect the environment. To focus on the needs of others is in the long-term interest of business. While he welcomed an increased interest in spirituality and meditation in the workplace, he said that developing positive, altruistic motivation is most beneficial.

The subsequent dialogue with business leaders gave rise to the following recommendations:

- Compared with five years ago, there is a definite trend, globally, of connecting business and ethics because reality is more interconnected than before. The need to broaden our perspective and consider the long-term consequences of business actions is becoming more relevant.

- Trade, investment, and technology are important for development but they are not enough: people should be educated to develop positive personal qualities such as compassion to boost their self-confidence, initiative, and creativity.

- We need to empower the full capacity of people working in business, which requires companies to develop a more human-friendly and inclusive culture.

- It is important to pay more attention to developing and promoting more female leaders in business, who have exactly the same leadership potential as male leaders.

- Positive change in organizations can only come about through leaders who are motivated by an authentic purpose and long-term commitment.

- We need to make changes in business education in order to develop leaders with emotional intelligence, who are governed by their heart as well as their head. They need to expand their attention beyond generating profits to combining wealth acquisition with creating increased societal well-being.

In April 2004 in Irvine, California, we gathered 115 international business leaders from our network to meet with the Dalai Lama. The meeting was convened in association with the University of California in Irvine and moderated by my cofounder and cochair of Spirit in Business, Anders Ferguson.

Anders Ferguson: Thank you, Your Holiness, for taking the time to continue this discussion about compassion or competition, and to discuss today the theme of designing an economy that works for everyone. There are 115 of us that have come here today; 50 people made the trip by plane, several people have come from Europe, others from South America, and they represent all areas of the business and academic communities. I think it's quite a tribute to everyone here that they are taking the time to look at such a big question. Your Holiness, would you care to make some remarks about the overall question of designing an economy that works for everyone?

The Dalai Lama: From some 40 years ago, I have had a keen interest in global issues. Of course, I'm not an expert, but as a Buddhist and as a monk, I pray for all sentient beings, wherever they are, and that includes the limitless galaxies of the universe. But on a practical level, this offers little help to our immediate concern, which is our own planet, isn't it? So, when I first went to Europe in 1973, I expressed the idea that we all needed some sense of global responsibility, a sense of universal responsibility, some kind of global ethics. This is not intended as a religious concept, but as something practical: we need to be thinking about the whole world because we are part of the world. If peace and prosperity remain in the world, we will all benefit. If the world faces more problems, including damage to the environment, then there is no hope for one's own bright future. Initially, I thought deeply and engaged in dialogues about issues of science and technology, where I could immediately see the important role of values such as compassion. But business and the economy were areas where I felt it was quite difficult to see the role of ethics and compassion. The connection didn't seem that obvious.

We have witnessed the implementation of socialist systems in the world. They sounded very nice. They said that these systems were in the interest of the majority: working-class people and needy people. But then time passed. The socialist economies weren't very successful, whereas capitalist, market-oriented economies were very successful. The latter had a dynamic force caused by a sense of competition that was lacking under socialism. So competition, as a feature of the capitalist system, is very necessary in order to stimulate individual initiatives; it seems that competition, sometimes even ruthless competition, is actually working! But then over time I began to see other things. Eventually, at the "Compassion or Competition" meeting in Holland in 1999, I met quite a number of businesspeople who showed a genuine interest in ethics and values. I spoke to some individuals who were working in big corporations, who were really showing an interest in social values and some even in meditation. But I'm not sure what was behind this interest. The idea that meditation and values may just be for better business initially shocked my mind. It seemed some people were interested in developing a sharp mind guided by ruthless competition! Of course, I don't know if this was the case.

> The idea that meditation and values may just be for better business initially shocked my mind. It seemed some people were interested in developing a sharp mind guided by ruthless competition!

There may well have been some other, more positive motive. In any case, I see more and more people now showing interest in spirituality.

Recently, I met a business consultant who works with many big companies, who understands that we Tibetans are at an initial stage of our own economic development, with unique conditions. He has given advice and training to Tibetan businesspeople, which I consider very important. Business has a considerable role to play in creating positive conditions in society. But there are also big corporations who disregard the social and development needs of disadvantaged people, and their actions have led to disaster. These problems come from a lack of social responsibility, from a lack of moral principle.

I am glad to see that in the business world, values and ethics are now more important. Many companies are very much concerned about creating and maintaining a positive image, and preserving their reputation. So, more and more companies are now showing some kind of moral responsibility; that's good.

I really appreciate that now, here today, there are many people who are connected with business, who actively work in the economy, and are showing concern about some of the drawbacks of modern business. We also notice the growing gap between rich and poor. We have seen that Marxism and the socialist system have generally failed; now, we can appreciate that the market economy, or the capitalist system, has some negative consequences too. For example, in modern China and also in Russia there are huge and growing gaps between rich and poor people. This is also the case in India and Brazil, other large developing countries. In America, the richest country in the world, we are witnessing the same thing. On one occasion, I observed that in Washington DC, the capital of the richest country of the world, there are also many poor people. Because of economic disparity between these poor families and others in the community, they are mentally unhappy. As a result, there is more violence and more crime.

Even among those from a rich family, there are also victims. Of course, if someone is hurt as a result of economic disparities within a community, the community is neither a healthy nor happy one. So, this gap between rich and poor must be reduced. Just thinking about profit, money, and earnings, and without hesitation exploiting things for your own benefit, that is not good. That must be reduced, it must stop.

> Self-discipline is not sacrificing one's own interest; in fact, it is aimed at protecting your long-term interest.

How can we do that? Not by regulation or law only. Some countries do that, but the effect is limited. Like any discipline, if discipline is imposed by force, in the long run it will not succeed. Instead, we need self-discipline, voluntary discipline, based on knowing the long-term consequences of our actions and then deliberately refraining from these actions. Knowing that even though there may be some immediate benefit, if we believe the long-term results are no good, we should stop these actions. Self-discipline is not sacrificing one's own interest; in fact, it is aimed at protecting your long-term interest.

Some companies do seem to operate like that. I like the case of airline companies. In airplanes the air hostesses always smile and, in India, for example, where the plane is very often delayed, sometimes one hour or more, they announce that they very much regret that delay. Actually, I don't think they have any regrets. So why do the hostesses say that? To bring more customers back. If they were to say, I don't care about the delay, the customers will complain and may choose another airline. Likewise, if you look at shopkeepers, although their main interest is profit, they know that the proper way to achieve this is not by force, but by smiling or some friendly attitude. This may, ultimately, be self-interested, but they have to show that they care about others. That's the human way.

> if you look at shopkeepers, although their main interest is profit, they know that the proper way to achieve this is not by force, but by smiling or some friendly attitude

In order to protect one's own long-term interest, including preserving a good image and bringing more satisfaction to customers and also to society, businesspeople are taking care of their employees and customers. Some are also paying serious attention to the long-term negative consequences of their actions, including those affecting the environment. In the long run, it is in the company's own self-interest to care about others; essentially, taking care of us all is in one's own interest. I think, through awareness of all of this, we can develop a certain discipline. I think that's possible.

> In the long run, it is in the company's own self-interest to care about others; essentially, taking care of us all is in one's own interest.

All these things are part of education. Through meetings like this one today, we can shed more light on the importance of moral ethics in the field of business. I am very happy to have this kind of meeting. So, thank you.

Anders Ferguson: In that spirit, Your Holiness, we'll ask the people in this room to bring their expertise forward. First we want to discuss the role of trade and market exchange in creating global prosperity and peace. Modern economics rests on the principle of free trade yet, despite much of the financial value that has been generated, it does not seem to work for everyone, given the persistence of poverty, social inequality, environmental degradation, and so on.

John Graham: Your Holiness, my mantra lately has been "trade causes peace." Of course, I'm not talking about the weapons or narcotics trade, I'm talking about the principle of free trade, through the mechanism of the market, which now has reached all over the world. In Europe, there is a rise of the concept of "fair trade" which ensures a fair and equitable distribution of wealth for those involved in the trade, avoiding issues such as human rights violation, exploitation at the bottom of the pyramid, and so on. My question for you is: does trade cause peace? Is trade good? Since you've had former lives, can you talk about the historical view as well?

The Dalai Lama: Most probably in my many past incarnations I was involved in a specialized field, not in trade. I think if I were to get involved in business, probably I'd end up creating more debts for the company, which indicates that in my previous life I was not familiar with how to make a profit! However, I do think that trade, particularly in modern times, is a very important channel for communication with other people. Generally speaking, trade is something that benefits the people involved—it is a mutually beneficial exchange. Through this exchange, people obtain a closer understanding of each other. They become friends, and that's the foundation of peace, isn't it? Personal contact is very, very, very, very crucial. And that is one way for the removal of fear or doubt, so I agree that in this way trade can help create peace. But if you use trade to secure other interests, then it is a different question, isn't it? In some case trade means exploitation. The stronger party may exploit the weaker one. And that creates more division, not more peace.

Barbara Krumsiek: Your Holiness, I'm grateful for this opportunity to be here with you. I run a company as a CEO that invests in companies that not only will deliver financial returns, but also operate in a way that is conscious of impacts on workers, on the environment, and on human rights issues. We call this socially responsible investing and my wish is that this type of investing was more common. It is not. It is a very small segment of the total financial industry, probably 2%. One hope is that in an economy that works for all, we'd have more awareness of this kind of investment strategy. How do we balance the short and the long term in a business sense to encourage more investment and commercial enterprises to think more long-term? Or can significant change only be done through individual ethics and will?

The Dalai Lama: That's a difficult question. Perhaps, while you are looking at the immediate benefits of a decision, you can also try to constantly analyze possible side-effects, considering the long-term consequences, which are often very hard to predict. I suggest you broaden your perspective. As a commercial enterprise, you have to take into account the need for immediate returns that the shareholders expect you to bring. But as a socially responsible business, the added dimension is to constantly bear in mind what kind of long-term impact your business activity will have overall. Within your enterprise, you could probably have a separate wing whose responsibility would be to really look into this long-term impact of the company's activity.

Barbara Krumsiek: If I may follow up, perhaps that could be something we could ask of the companies we invest in. Where is that questioning going on in their enterprise? So then the longer-term thinking will ripple out.

The Dalai Lama: One thing I notice now is that, even within the business world, more and more people seem to be at least talking about the ethical dimensions of business, and there is also a greater awareness of the global dimension of business activity. It is just the beginning, but I think the trend seems to be moving in a positive direction. So just like you, others should put more effort into this trend. That would be very good.

4.1 The role of technology: cure or curse?

Brian Arthur: We have had an extraordinarily successful 200–300 years in the Western economy due to the discovery of new technologies—everything from better plows to textile machinery to cures for diseases and ways to diagnose disease. So, in a sense, technology has given us an enormous amount of things, including much longer and healthier lives. What can we expect now? We are starting to intermingle technological advances with areas that are very powerful, but we're not quite sure they will be beneficial to us. My question is: how should we think about technology in the long run? We're entering a very new era with enormous possible benefits. But things can possibly go wrong, and there are some things, such as genetic technologies, that may be blurring our sense of what it means to be human.

The Dalai Lama: My personal feeling is that if you look at the evolution of human society in a kind of broad sweep, certainly there is a natural evolutionary process, and progress which would have occurred in some sense anyway. Of course, there is a level of human effort and human skill and human knowledge, particularly combined with human technology, which has brought about a much faster pace of development. But even with that kind of technological development there are certain constraints posed by natural laws in the physical world. Even in the new age of genetic technology, this pattern will probably remain the same. But one thing that I think is especially important in this new era is to be even more vigilant about possible long-term negative consequences of this technological power.

I believe that technology, like any human activity, is connected to the benefit of humanity. If we keep in mind the benefits for humanity, we can act accordingly and experiment with these things. Of course, today there are many things that we are using for the first time, so the longer-term consequences are not yet known. For some people, if they are warned about possible future consequences, they feel this is unnatural, this is too extreme. But after one century, maybe that negative consequence has come to pass and has become something normal. You see, I feel possible negative consequences are there by nature, so unwanted things can happen. Ultimately you have to accept the path of natural law. In the meantime we can use our ability to investigate possible future effects to improve our understanding of a particular aspect of nature and help us make wise decisions.

4.2 The global responsibility of business

Mark Thompson: Thank you very much for being here, Your Holiness. I'm here as a messenger today on behalf on the rest of the world, especially those in developing countries, who tend to face much larger challenges than we do in the West. First, a recent incident that I feel is significant for our discussion.

When I was in Zimbabwe, I met a little girl named Shona, who was five years old. She had been orphaned a year earlier, losing her parents during violence in Zimbabwe. I went to visit her classroom and her teacher was talking about leaders for peace, including Mandela, Carter, and His Holiness. I spoke a little about some of my work with Jimmy Carter and Mr. Mandela and said that I looked forward someday with great expectation to seeing you, Your Holiness, at some point in the future. At that point this little girl, Shona, sprinted to the front of the classroom, tapped me and pulled me down until I was on just one knee so she could look me straight in the eye. She said, "I know all about the Dalai Lama. He brings a message of peace and I want you to give him this gift." I said, "Shona, I don't know if I would ever have an opportunity to deliver this gift and frankly you shouldn't be giving us anything. We should be showering you with gifts." She thought for a moment and said, "That sounds like a really good idea." I was still down on one knee at this point and finally I looked at what she had given me, which was a Shona sculpture. This is wonderful art that you find throughout Africa, particularly east Africa and Zimbabwe; since this little girl's name was Shona, I imagine that this was not an accident. She said that her mother used to tell her that her love and her heart were stronger than this stone, and that anyone who commits their life to peace should have this. I wanted to give it to you today—from Shona.

The Dalai Lama: Thank you. I would like to give a scarf for the girl, for Shona, that you have to deliver. [His Holiness carefully knots the scarf and explains what he is doing]. This knot symbolizes a bond and a friendship.

Mark Thompson: Thank you so very much. Now you've raised a question I must leave with you. It is very difficult and dangerous to return to Zimbabwe at this time, particularly as an American business person and journalist. I would appreciate, on behalf of all of us here, how you would counsel us to think about the best way that we can form a better bridge, a more sustainable and consistent form of support to people in countries where American businesspeople are more unwelcome than ever before?

The Dalai Lama: I don't know. It's a difficult question. But if we divide the world into different parts and treat them differently, whether it is Asia, the Middle East, or Russia, then we lose a great potential. Now as far as Africa is concerned, it is a huge continent. No doubt it has real potential. But unfortunately, I think people there generally seem to lack self-confidence. I have

From left to right: Anders Ferguson, Barbara Krumsiek, Bill George, the Dalai Lama, translator Thubten Jinpa, and Brian Arthur.

been to South Africa twice. One occasion I visited the Soweto township. I went to visit one family, where I learned something important. The family members and friends all joined the meeting. One was introduced to me as a teacher and we talked about Africa. As you know, South Africa recently became a democratic country, where legally speaking white and black people have equal rights. However, the teacher told me that this is just according to the Constitution; only on paper are they equal. I told him, this is not sufficient; the people must also work hard with self-confidence. Then he responded: "We cannot be equal with white people because our brains are inferior." When I heard that I felt very, very sad. And I argued with him: "That's incorrect, that's not the case, it is only your imagination!" Then I explained about my own situation. Sometimes some Chinese consider Tibetans as an inferior ethnic group. But since we became refugees in a free society as a result of our own hard work, we have shown we are equal to Chinese and others. In other words, if the opportunity is there, there are no differences. After my lengthy explanation, he said that he was now convinced we have the same potential. At that moment, I really felt a tremendous relief. At least one person, a teacher moreover, had gained a new sort of self-confidence, a new spirit. I felt happy that I had contributed to the transformation of one person's mental attitude. This is one example of what we can do.

On another occasion, I visited Gabon. There I noticed that within the country there was another big gap, between the Western educated elite and the Muslims. With such a big gap, these societies cannot transform. In

India, I used to tell my Indian friends, "Real transformation of India must take place in rural areas, in farmland areas. Then India will transform." Developing just a few cities is not sufficient. This applies to China as well. I remember that in 1955 I was in China for about ten months. On one occasion, as part of my visit, we toured the city of Shanghai and met the mayor. One evening he told me that he is not interested in the further development of Shanghai. His main interest was the rural population in the country. I remember this conversation very clearly. My main point is that we need to educate the large masses of poor people and give them self-confidence. You can give them technology or financial assistance, but this will not be enough. It must also come from them. They need to develop their own initiatives. In that way and through more education or more training, and then through earnest effort, they can gain self-confidence. This is my feeling. And I think this applies to all human beings; if you really treat them with sincerity and care, they will respond. This even applies to animals. They will respond to care and attention.

4.3 Leadership and organizational change

Bill George: I want to shift the conversation to a subject that has been my passion for a long time. As you are well aware, in recent years many corporations have failed. In the process, so much economic value and so many jobs and lives have been destroyed that our media has branded the leaders of these failed companies as bad leaders. However, in studying them closely myself, I've come to the conclusion that these people didn't set out to do evil things. Most lost their way *en route* to the top by seeking external gratification—money, power, titles, status, and prestige. What the leaders of many failed corporations never developed in all that rush toward success was what you might call an inner life—they failed to cultivate their hearts. As a result of this, one of the things we're trying to do with talented young leaders today is to start them early on a path where they understand that their lifelong development as leaders is parallel to the development of their inner life and the cultivation of their hearts. If you agree with this premise, what would you tell young

> What the leaders of many failed corporations never developed in all that rush toward success was what you might call an inner life—they failed to cultivate their hearts.

leaders to do today to prepare themselves for their significant future leadership responsibilities?

The Dalai Lama: My view is that it is not simply a moral issue. I believe, whether it's in the economy, or politics, or education, we need to carry out all human activities with human feeling, because everything in reality is interconnected. All the books young leaders read about politics, economy, religion, education, and so on—even if they address specific topics—are all meant for humanity as a whole. Taking the holistic view, even in your specific field of business and the economy, is best. Economics means dealing with humanity, and the situation of humanity changes century by century, year by year. Once they understand that reality is constantly changing and that everything is interconnected, leaders begin to realize that they have to keep the consequences of all their actions always in mind.

> Once they understand that reality is constantly changing and that everything is interconnected, leaders begin to realize that they have to keep the consequences of all their actions always in mind.

Although a leader may get some immediate economic profit in a particular case, there may be other important side-effects that weren't considered. The broad-minded perspective and the more holistic view, which can see reality in its totality and understand the longer-term effects, is essential for leaders today.

For example, I'm a religious person, a religious practitioner. I am a Buddhist but I also respond to non-Buddhist and nonreligious people who ask me for teachings and explanations. When I am explaining something, I try to keep in my mind if my main audience is Buddhist or not, religious or not, and then I adjust my teachings accordingly. If I only look at my Buddhist interests, I may actually be in conflict with my other aim of promoting religious harmony. So, although I'm engaged in

> Reality is so complex, so interconnected; we all need a very broad perspective.

explaining Buddhism to certain people, at the same time I have to keep in my mind the implications for other, non-Buddhist people. If you just think about your own religion and nothing else, then—although your motivation may be sincere—there could be some undesired consequences. I know that some religious people, Christians for instance, speak about their religion very sincerely and with great enthusiasm, but if they only speak from their own specific religious perspective there can be negative consequences. So, that is my view with regard to business leaders as well. Reality is so complex, so interconnected; we all need a very broad perspective.

Peter Miscovich: I work for a large consulting organization and we help organizations transform themselves so that they become better workplaces, perform better, and create more value. My question is: how do we expand and have these new concepts of holistic thinking, long-term views, and compassionate motivation embedded into global organizations? What do you recommend in terms of organizational transformation to help people embrace these new concepts?

The Dalai Lama: Would any of the panelists like to respond to that? At this point, I think you, the other panelists, will have more experience.

Barbara Krumsiek: I would just quickly say that I think there are two elements to transformation. There's the leadership and then there is every person in the organization. I don't think transformation is possible without both of those elements being in alignment. Leaders have to be able to set the tone and be able to inspire every individual—in some senses, everyone has to buy into the transformation. I think the harsh reality may be, as His Holiness has said, that if some people will not buy into the transformation, then maybe they don't belong in the organization.

Bill George: I agree with that, Barbara. I think organizational transformation has to be throughout an organization. You can't take a factory within a company or a division within a corporation and transform it. The key is a common mission, people coming together around a common sense of purpose. And it can't just be a set of words. It has to have real meaning that's felt deep inside, and a common set of values so that people can develop trust. And then it has to be tested in the crucible. It has to go through the refiner's fire. In other words, it has to be upheld even in difficult times and when things are not going well. And if the organization stays true to its purpose and its values in that difficult time, for example, when the market share is going down, the organization is losing business, the economy is bad, and some key people quit, that will be the test. And, if the leadership then goes

> The key is a common mission, people coming together around a common sense of purpose. And it can't just be a set of words. It has to have real meaning that's felt deep inside, and a common set of values so that people can develop trust. And then it has to be tested in the crucible.

in a different direction, and says, "Oh, now we can't stay true to our purpose and values, we have to do this because things are tough," then the transformation will be lost forever. You are better off not to start the transformation in the first place. So, the leadership has to be fully committed, knowing they will be there when times are difficult; because if they're not, then there will be just a sense of cynicism across the organization. So many leaders try to start these efforts because they think it is a good thing to do, but they haven't really thought through the whole process and aren't prepared to stay committed through difficult times.

Finally, I think we have to recognize that leadership in a transforming organization must exist at all levels. People on the production line, people serving the customers, people in the finance department, everyone has to lead and the leader has to be out with the people. If the leader is in Wall Street at security analyst conferences, it will all fall apart. And then, the organization needs to say very clearly what it stands for to the outside world so that it will have a buffer—an understanding on the part of people, as Barbara Krumsiek says: "This is what we stand for. And if we go through difficult times, we're asking you to stay with us because we've stayed true to our mission and our values." So, that's the only way you can have long-term transformation.

It takes about five years. If you're interested in a six-month transformation, forget it. You're just trying to impress the stock market and it won't work. It takes five years because you have to go through difficult times before people believe you. So, if you're serious about it, as for example IBM has been very serious about it, it's going to take at least five years, and in their case, ten. And that's why so many organizations who have attempted have failed so many times because they aren't really serious about it. It has to be a transformation committed to and with people. It can't just be an external transformation, the mission statement on the wall. It has to be something that goes on inside people so there's a belief there. And that's a transformation of consciousness. And if there's not that transformation, then it can't work. People have to believe that their leadership really cares about this coming together around a common purpose, a set of beliefs, and a set of values.

> It can't just be an external transformation, the mission statement on the wall. It has to be something that goes on inside people so there's a belief there.

4.4 The role of women in business

Karen Buckley: My question comes from a point of view that I have at this time that the voices and wisdom of women leaders are ever more important in our world. I'm wondering if you could speak to that from your travels and your experience and your understanding. What is the role of women as you see it, and particularly those in leadership positions over this next decade?

The Dalai Lama: I believe basically that males and females, except some differences on a physical level, are very much the same, with the same brain and the same degree of intelligence. In some cases, I think females are sharper. In any case, I think that as far as loving kindness is concerned and the tendency to care, females are more experienced. But essentially, they are equal. However, it seems that today among top leadership, be it of religious, political, or business institutions, the majority is male. Perhaps this is because for a long period the women's self-confidence was a little neglected. Now, I think things are changing. I think that discrimination between men and women is a backward sort of thinking. In ancient times, unlike the modern world, the body strength was the main attribute. So people with the stronger bodies became dominant. Now, in modern times, obviously some of the leaders are physically small, but their brains are very smart. Just with a small body they can lead a large nation or an organization. Therefore, the physical aspect is secondary, it is not important anymore. In business, men and women are now equal. So I think discrimination is a leftover from some sort of old habit.

> I believe basically that males and females, except some differences on a physical level, are very much the same, with the same brain and the same degree of intelligence.

The important point is for women to have self-confidence, to work hard, and then they will be equal with males in society. But some feminists are a little extreme. It seems as if some of the feminists wish that Buddha and Jesus Christ should be female. That's I think a little extreme. After all, males need females, females need males—they must go together.

Barbara Krumsiek: I want to hold onto self-confidence, work hard, and be equal. That's what I heard from your Holiness and I agree with that. But I also do think there are some structural impediments that perhaps followed historical or ancient differences and have resulted in some actual structural barriers. An expression that has been used a lot in business is the "glass ceiling," getting close to the top but not getting through. So, I know that through our own efforts at Calvert Group, the dialogue with companies we invest in is focused on board representation, and we have successfully influenced several companies. We have written to 650 of them and requested that every time they search for board members, these searches should include women in the pool of candidates, just as they would in their senior executive searches. It isn't sufficient to have one woman and 12 men; every search should include both women and minorities. We've drafted language that boards can adopt for their nominating committees so that their structure can be more inclusive of women and get the equality that we think is appropriate.

Correct me if I'm wrong, but one of the things we see if we look across professional schools is that the majority of law students and medical students are now women. That's not the case in business schools. I think some extra work is needed in the field of business.

The big issue with business is somewhat different. Today we're in an era of two-career families and we need to reshape the corporate workplace to reflect this reality. And we haven't done that yet. Yes, I agree that women in significant positions throughout an organization bring an added dimension to the workplace. I think that's a very healthy environment. But the number one issue for all the young people I know in their 20s and 30s is: can I have a successful career and a successful home life, family life, and personal life? That's not just women speaking. That's men too. So many careers are still set up on the old male model of career development. We need a new way of thinking about our careers and how to develop models that can be integrated into successful corporations so that they can be competitive in the world. I think in the future both men and women will want to go to work for companies that offer that kind of an environment. But it's a big issue right now and we haven't got these models developed yet. We need to focus on these developments now.

The Dalai Lama: Yes, I think this is very important, because it would then be applied to all the employees who would seek employment with the company.

Barbara Krumsiek: Exactly. Many corporations are saying, particularly to professional women, we expect employees to work 80 hours a week. And I know women who are trying to raise two children and work 80 hours, and it's just not possible. Unless you have a husband at home taking care of the home tasks, I think we need to rethink the balance in our lives. And frankly, we'll get better leaders, and we'll have healthier organizations if we do that, because in the end, people will have much more loyalty to the organization, much more commitment, and you won't have these high turnovers and you won't be faced with the difficulties of women exiting corporate positions.

4.5 The role of management education

David Cooperrider: Your Holiness, it's wonderful to be with you again and the message in your book *Ethics for the New Millennium*[1] is very important and has inspired me in your call for a radical reorientation toward the "other"—those who we regard as strangers, who are different from "us." Last time I was with you, we were in Jerusalem with a small group of Muslims, Christians, Jews, and Buddhists and I felt that radical reorientation when you took us all with great respect to the Jewish wall, and we prayed and meditated together, and then to Christ's tomb and we prayed and meditated together, and then to the Dome of the Rock and we again prayed and meditated together. So, I thank you for your leadership and example for that radical reorientation to the other. I also want to thank you for this kind of conversation with business leaders. It's so important today. I'm a management school professor and I think we have a real opportunity to take action today, to move from a dialogue like this one and to take some action.

I don't know if you know it, but in the United States over a million students a year are trained in our management schools. And these are the people that are going to be making billions of decisions every day on behalf of the planet. The Aspen Institute, a great leadership development institute, wants to create a positive revolution in management schools, which I think is what we need, and they're calling all of us to think about reshaping education in management schools to create a tipping point of change toward a radical reorientation of the other. If you were part of that

> What does the management school of the future look like? If anything imaginable was possible in terms of our curriculum, in terms of our teaching, how do we really create leaders with hearts?

1   Dalai Lama (2000). *Ethics for the New Millennium*. New York, NY: Riverhead Books.

committee advising them, what does the management school of the future look like? If anything imaginable was possible in terms of our curriculum, in terms of our teaching, how do we really create leaders with hearts, the kind that create this spiritual shift that you're talking about?

The Dalai Lama: I'm not an expert on management. To be frank, I feel I myself am not a good manager. If I were to manage something, it would end up in chaos. If you come to my room, everything is very untidy. [Laughter]

You have more experience and I leave it to you to make suggestions as to what should be improved in business education. But I feel this sort of growing enthusiasm for wanting something to change, based on the realization that something is not satisfactory, is very good. Usually we take things for granted and go on according to tradition, in spite of some drawbacks. That is a real failure, which is preventing progress. But this enthusiasm, saying that something is wrong, is the first step to lead toward transformation or some positive change. We then need more discussion, more research, and more experimentation. Then we will find answers.

> We need more discussion, more research, and more experimentation.

Brian Arthur: I think we've taught students how to do managerial finance and we've taught them economics and so on. We've taught them, particularly, to look out for the bottom line, to make profits; but there's one thing I think is worth teaching students and that is to ask a simple question: how can I serve? And when I join a company or when I run a company, how can that company serve? What I see is missing today, which may not have been missing a hundred years ago or more, is just a simple pride in what companies produce.

I had the good fortune to know a wonderful man, Ernesto Illy, who produces coffee; he's an Italian coffeemaker, a really wonderful man, and he's the CEO of a successful company.[2] Best coffee, I believe, around. But what I noticed in him that seems to be missing in many is a simple pride. He knew every bush, every tree in Ethiopia where the coffee came from. He cared about the coffee. He cared about the product. So, it wasn't, "Let's think about the customer; let's think about the community." It was a simple thing. "My goodness, I am so proud to be producing this and isn't this coffee

2   For further information, see https://en.wikipedia.org/wiki/Ernesto_Illy.

wonderful?" So, if we can teach our students simply to serve and take pride in what they do, that would be great.

I was in China about four years ago, standing in line for a flight and I started a conversation with the fellow in front of me. I asked him what he did and he told me he was a sales manager for ABB. He asked me what I did and I said I teach international business. And he said, "There's no such thing as international business. There's only interpersonal business." And that was a very wise answer. But the point is that the people in business schools may need to work hard on this too.

> There's no such thing as international business. There's only interpersonal business.

For the past 15 years, we have been selecting MBA students for their IQ and training them about how to manage with their heads. I think that we need to select people who have leadership capacity as evidenced by their emotional intelligence and their hearts and then teach them how to lead with their hearts, as well as their heads. We'll have far better people coming into the business world if we can engage the whole person. Much like in medicine, we teach doctors how to treat disease rather than how to heal people as a whole person.

> We need to select people who have leadership capacity as evidenced by their emotional intelligence and their hearts and then teach them how to lead with their hearts, as well as their heads.

4.6 What can you do as an individual?

Terry Pearce: I wanted to ask a very simple question. We have been discussing some very complex issues, but my thought is very personal. Given our common objective here, how should I live my life? What can I do as a individual person who happens to work in business?

The Dalai Lama: I think in order to answer that question I should first know what your background is. And what is your usual habit or way of thinking? Even then, I do not have the full answer. In any case, generally we, each individual, want a happy life. It is something we deserve—to be happy. That is our right. But our respective circumstances are different. We humans are social animals. If you are a businessman, then, of course, today you operate in a world where national boundaries are not important anymore. You know that everything is interconnected. Under these circumstances, I believe that a compassionate attitude toward your friends and toward strangers that you meet, and even toward your own enemy, is very useful. This compassionate attitude really gives you more inner strength, more inner peace. I think this will be of tremendous help to boosting your potential when facing the challenge of working globally in a changing world.

Likewise, if you have a family, I would suggest that you bring up your children in a more compassionate way, not only focusing on giving them a good education but also on developing their spiritual qualities, such as a compassionate attitude or kindness. Then you can learn to extend this compassionate attitude to your neighbors, and beyond. I think every single one of us can do this. The long-term transformation of a society must be carried out by individuals. Transformation of a society by government or law is difficult. It has to come from individuals and their families. Within a family, someone must take the initiative. The transformation of other family members can then happen, and then the neighbors, and so on. It is natural that within families and among families there are some disagreements. Some will enjoy music, but some others may feel the music is too noisy. That is natural; it is part of human life. In spite of these differences or conflicts, we are still part of the same family. That reality is not changed by the conflict. Though we may have different individual interests, we do share a common interest as members of the same family or community or even humanity. If we consider our common interests as the top priority, the little differences are secondary and we can overcome them.

> If we consider our common interests as the top priority, the little differences are secondary and we can overcome them.

But we usually approach it in the opposite way. Our little differences become the top priority. We forget about the sameness of our human nature. I feel that this also applies to the religious field. We should consider our humanity as the top priority. Religious faith and religious differences are secondary. Again, we often do the opposite. But I think that considering our common human interests is the proper way to promulgate human compassion and promote world peace.

Betsy Denham: Your Holiness, how can we as an audience today take the words that you're teaching us and become change agents, either locally in our own organizations, or carry the message beyond? So it's more than just today, but something we can do 30 days from now so it's measurable. And at that point, how will we know that we've actually seen a change from our actions?

The Dalai Lama: Of course, it depends very much on the individual circumstances. Among the participants here, there might be individuals who will have the capacity and also the opportunity to immediately implement many of the suggestions that have come up through the discussion here. However, there will be individuals who, even though they share the same views, may not have the ability to immediately implement them.

I myself am included in the second category. I share all the views that have come up here, but the only practical thing I could do is to make prayers that these things become a reality. Among the Tibetan community, business activity is still quite poor and, even among the Tibetans who are living in America, I have hoped that there would be some millionaires emerging, but so far no luck. [Laughter]

However, many of the underlying ethical questions that we have been discussing are not necessarily confined to the areas of business and the economy, but they are something that needs to be immediately applied in the context of one's own individual life, family life, and community life as well. So, in the case of your second point about when will we know when we have been successful in our change and transformation, particularly in the domain of business and the economy, I think that probably the media will tell us.

Anders Ferguson: Your Holiness, we're coming to a close. Do you have some final comments that you would like to give us at this point? Some final remarks please?

The Dalai Lama: I have really enjoyed our conversation. I appreciate the kind of enthusiasm you all have for addressing the inadequacy of how things are currently. Seeing something, and seeing that some improvement is needed—that's a very good start. I very much appreciate this meeting. This is a very encouraging sign and also, of course, it is a step forward.

Part 3:
Leadership for a sustainable world

5
Shared purpose in business

This chapter describes how I return to the world of (big) business, as I had come to see that this is where I could have the biggest impact for positive change. From my broad inquiry into the societal dimension of business and economic development, my attention shifts to helping organizations to find their societal purpose. I specialize in leadership development, executive coaching, and organizational change, and then work with a few leading companies, which gives me hands-on experience with the challenges involved in organizational change for societal benefit—some encouraging, some discouraging. While some companies remain very dependent on the system of short-term capitalism, others are finding ways to transform themselves into long-term value generators for society.

In this phase I am able to connect my purpose to helping companies find their larger societal purpose. I define this experience as "finding shared purpose"—realizing the shared connection between business, society, and the natural environment. Here I find new questions: how to develop purposeful leaders? How can they transform the purpose and culture of business? If the overall incentive structure is not conducive to the purpose of the company, what can leaders do?

The chapter ends with the story of the sudden disruptive change in China and the collapse of banks in the emerging financial crisis. This formed the backdrop to the third public dialogue with the Dalai Lama, which focused on leadership for the modern age.

5.1 Emotional intelligence and system change

At one of our conferences I met Dr. Annie McKee, who had just co-authored a book called *Primal Leadership*.[1] The lead author was Daniel Goleman who is famous for introducing the concept of emotional intelligence. The book, which integrated and expanded the work on emotional intelligence into that of leadership, became an instant bestseller. What readers liked about *Primal Leadership* was the fact that leadership theory, for the first time in history, was firmly rooted in the biology of the brain and human emotions. This scientific field of neuroscience had discovered that the brain is "plastic" in the sense that it is changeable and trainable. As a result, leadership was no longer perceived as merely a personality or character trait that you would either possess or not possess—end of story. Significantly, it had now been demonstrated that there were no "born leaders," and leadership was not a formal function in an organization, regardless of any leadership ability. Rather, leadership could be seen as a science of how the brain works in relationship to other people. This was the science of emotional intelligence, which was more primal in human beings than cognitive intelligence— hence the title *Primal Leadership*. Since brains can be trained, leadership could be regarded as a practice of training the brain. That is, by training the brains of leaders the brains of followers will also be changed.

Aside from being a successful author, Annie was also a coach who was working with senior business leaders. Because of the book's success, the demand for her coaching work exploded as multinational companies approached her with the request to design and deliver company-wide leadership programs. So Annie was on the lookout for people who could help her. When she learned that I had been a banker, she invited me to work with her on an assignment for a major international bank, HSBC. This would mean that I become a leadership coach. She said that aside from knowing how to speak the language of business, I needed to be versed in emotional intelligence and systems thinking—the emerging science of how complex systems (such as global companies) work.

Interestingly, I had met Daniel Goleman in his role as keynote speaker at our first Spirit in Business (SiB) conference in 2002 and I was aware that he was a meditator who had developed much of his work on EI (as the field of emotional intelligence became known) from his training in Buddhism and

---

1   Goleman, D., McKee, A., & Boyatzis, R. (2002). *Primal Leadership: Leading with Emotional Intelligence*. Boston, MA: Harvard Business Press.

other contemplative traditions, and had regular dialogues with the Dalai Lama on these topics.[2] Buddhist practice can be divided into two elements, wisdom and compassion, which are considered the two essential features of spiritual development. They are like the two wings that a bird needs in order to fly. Wisdom provides insight into the reality of transience and inter-dependence, while compassion gives you a sense of purpose—a path away from selfishness, which is considered a major cause of suffering. You could say that EI represents the compassion side of leadership, whereas systems thinking represents the wisdom side.

As I had familiarity with training in both business and Buddhism, I jumped at the opportunity to join Annie McKee and her colleagues and became an EI-based leadership coach and consultant at the Teleos Leadership Institute in Philadelphia. I was keen to reconnect to the corporate world in a profes-sional sense and learn about new developments. I was especially interested in trends such as corporate responsibility and sustainability, which for me were part of the larger quest of creating a more compassionate form of capi-talism. The SiB dialogues had taught me that all system change (not only technical or structural change) ultimately requires leadership. A new type of leadership that unlocked a human's full potential for the benefit of trans-forming the larger system was now required. I wanted to explore this type of new leadership.

5.2 Positive psychology in leadership

The founders of Teleos, Dr. Annie McKee and Dr. Frances Johnston, served on the faculty of University of Pennsylvania, from which the field of posi-tive psychology, spearheaded by Dr. Martin Seligman, was rapidly emerg-ing. As we saw in Chapter 3, Seligman sought to redress the orientation of mainstream psychology toward human shortcomings and malfunctioning, instead focusing his work on understanding how humans can develop posi-tive qualities that enhance their well-being and happiness. For some strange reason, psychology had ignored the positive potential of people, which is

2   For a record of these dialogues, see Goleman, D. (1997). *Healing Emotions: Con-versations with the Dalai Lama on Mindfulness, Emotions, and Health*. Boston, MA: Shambhala; and Goleman, D. (2003). *Destructive Emotions: How Can We Overcome Them? A Scientific Dialogue with the Dalai Lama*. New York, NY: Ban-tam Dell.

so obviously relevant to the development of leaders. How do people grow? How do people keep growing when they take on larger responsibilities in life? How can they grow in such a way that they become more effective, and retain or even improve their sense of well-being? How do performance and happiness relate? It was a great opportunity for me to learn from this emerging field of positive psychology and help translate it into leadership psychology and organizational change, both in theory and in practice with major companies.

It would mark a new phase in my work, bringing me closer to what I felt to be my purpose, which I started to define in terms of bringing humanity back into business and economics. I had left banking because I felt that it had lost its direction and I no longer believed its purpose of serving human- ity. I had discovered that I could not flourish in an organization without a purpose, and I had rediscovered my own sense of purpose in working in Bhutan and Tibet and with the many inspiring business initiatives that were manifesting in the SiB world. But having had the experience of working in and with "big business," and having experienced firsthand the potential impact on people's lives that big business holds, I felt drawn to transform big business into a force for good. How to awaken purpose in business— that was to become my purpose. In a sense, I discovered a "shared purpose" with business.

It represented the third stage of the journey of purpose: "finding shared purpose." While the first two stages described in Chapters 1 and 3 respec- tively could be viewed as being rather introspective, as they entailed an inward process of exploration and discovery of intrinsic motivational drives, the third stage was directed more outwardly and could be connected with the needs of others. I discovered how I could serve the needs of leaders who were looking for a purpose in their work that is connected to the larger world. My motivation to make a difference got connected to the needs of others. In that way a single purpose became a shared purpose, with new motivational energy as result.

5.3 Executive coaching for bankers

My first client was the global bank HSBC, and my job was to coach a few of their top executives in becoming more effective as leaders. It was strange to suddenly find myself back wearing a pin-striped business suit and be in a skyscraper in the city of London, a world that I had said good-bye to a few years earlier. I was struck by how the worldview of the bankers differed from that of people I had worked with since I left banking. The bankers were all highly educated people with global experience, but they had learned to see the world through the lens of big numbers, big deals, and big market shares, with fast and restless minds. Most of them suffered from relationship problems: both within their teams and with their families. Their work totally absorbed them. Hardly anyone worked less than 12 hours a day, which meant that they would arrive home too late to see their young children and meaningfully connect to their spouse. I had been there and knew how it felt. They were locked into a career with the aim of making enough money to retire early, and then to make up for the time they had lost. But, for some, this was not working.

Rick, in his late 30s, was at a loss as to how to appease his wife for the long hours that he worked. When he was at home at the weekend, he was often too tired to be a good husband or father. Back at work on Monday, he felt guilty for having been stressed out and bad-tempered due to tensions within his team. As team leader, he felt he had to be the pace-setter for his team members. His boss was constantly pushing him to deliver higher margins and turnover. I could see in Bill's eyes that he was on the verge of burnout. Rick was very smart but he lacked *emotional* intelligence—he did not know how to handle his emotions and to create an environment where he felt emotionally balanced. He felt stuck in a vicious cycle of work, fatigue, and distress. How could I help?

I started by helping him to set his priorities. Did he love his wife? Yes. Did he love his two sons? Yes. Did he consider their happiness more important than making money? Well, it took a bit longer for Rick to acknowledge that but, when he finally got there, he could start developing a new work schedule. He decided that for one day in the week he would take his kids to school and get home early, and plan a monthly private dinner date with his wife.

In the process of setting his priorities, he saw how he could stand up to his boss and improve the relationships within his team. As the team relationships improved, Rick started to feel supported by his team rather than being at the mercy of the team. He discovered that taking a stance on work–life balance actually gave him more self-confidence. He had something to stand for. It also strengthened his credibility among colleagues, most of whom were struggling with the same issues. In the end, he managed to improve his private life while still delivering enough financial return.

5.4 Unilever: examples of sustainable business transformation

A particularly interesting client was Unilever, the Anglo-Dutch fast-moving consumer goods giant, which had embraced CSR and sustainability as a strategic opportunity and which simultaneously invested in leadership development. For example, it had organized leadership trips to locations such as Costa Rica, China, and the Sahara in order to awaken a new vision for its future. Its board advocated sustainability from a straightforward business perspective: since consumers are part of communities in a larger social and ecological context, the company's success depends on the success, happiness, and the health of those communities. Consumers can only grow to the extent that their communities grow. Therefore, it made strategic business sense for Unilever to help improve global communities.

I was retained as a coach/consultant at a time when the company was in the process of restructuring its global organization around the concept of the value chain, which was to be led by categories of the major products. It was a highly political process; country organizations resisted the power transfer to a global category level. However, a capital efficiency requirement forced Unilever to dismantle the traditional country-based organization and adopt a global value chain-based structure, which helped the firm to operate with more alignment to globalized markets. In this way, Unilever became better positioned to serve the longer-term and latent needs of consumers and their communities. Value creation for consumers rather than production targets set by factories was to drive the business.

Though this transition was not motivated by sustainability concerns, it did in fact help the company to move its sustainability agenda significantly forward. It expanded the firm's sustainability focus from the supply chain to the entire value chain, including the consumers.

The supply chain focus had led to various commendable CSR initiatives, such as minimizing the ecological footprint, optimizing resource efficiency, and respect for human rights. But there had been a tendency to treat CSR and sustainability as an add-on, motivated by risk and reputation control, not as a necessary part of the core business. With the shift to the value chain perspective, sustainability became an inevitable trend, both impacting and driving consumer and community needs.

Unilever believed that, at one point, consumers would demand sustainable products and they would be prepared to pay for them. This value chain transformation process went hand in hand with investments in the generative capacity of Unilever's own people, because the sustainability value chain approach required different mind-sets and attitudes from its managers. Andre van Heemstra, the board director responsible for human resources, summarized this approach in a meeting with me: "CSR without HR is PR." I considered this a profound statement that I had not yet heard before in business. It was clear to the Unilever board that sustainability was not just a duty toward external stakeholders but an opportunity to educate and unify internal stakeholders. Because CSR was considered to be the "right thing" to do, it could help boost the work morale of employees. People were proud and happy to work for a brand that would take care of their clients and suppliers. And so sustainability became a win–win for Unilever: it helped position the firm on a new value creation trajectory while enhancing its own workforce capacity.[3]

As I was working with some of their top teams and visiting several of their companies across the world as this transformation was taking place, I started to sense the potential of such change. I learned most from working with Unilever's company in India, Hindustan Level, particularly with the case of the promotion of the soap brand Lifebuoy.

As one of Unilever's fastest-growing global brands, Lifebuoy is a great example of how integrating a societal purpose is also good for business. Lifebuoy was launched in the U.K. in 1894 and had championed a message of health through hygiene for more than a century. The Lifebuoy brand aims to make a difference in people's day-to-day lives by selling soap and encouraging hand-washing. In developed countries, modern innovations such as sewerage and piped water supplies, together with the widespread adoption of soap, have helped to reduce the incidence of infectious disease and reduce mortality rates from infection to 5% of all deaths. However, in Africa, 65% of deaths are due to infections, while the figure is 35% in Asia, including India. Interestingly, among all changes that are needed, hand-washing is shown to be the most cost-effective means of preventing infection and saving lives. By washing their hands with soap five times a day, children can be saved from diarrheal deaths.

The Lifebuoy managers in India asked themselves: "If hand-washing with soap is one of the most important preventative measures against disease, why is it not universally practiced in India?" They explored hygiene practices through investigating the various motives for (not) hand-washing.

3   Mirvis, P. (2011). Unilever's drive for sustainability and CSR—changing the game. In S.A. Mohrman, & A.B. Shani (Eds.), *Organizing for Sustainability Volume 1* (pp. 41-72). Bingley: Emerald Group Publishing Limited.

Unilever then created partnerships with the Indian government, NGOs such as Oxfam, UNICEF, and the Red Cross, local communities, and women's groups. Collectively they created Project Shakti, which employed women in local areas to educate communities and families on the benefits of hand-washing, and which also allowed Lifebuoy and its partners access to rural areas. Not only did these thousands of women become a new distribution channel for Lifebuoy, they were empowered through earning an income. Moreover, Lifebuoy's sales in India were boosted and, at the same time, there was a marked reduction in the rate of children's deaths. Children had 25% fewer episodes of diarrhea, 15% fewer incidents of acute respiratory infections, and 46% fewer eye infections. Children also had a significant reduction in the number of days of school absence due to illness.

This was a true win–win for both Unilever and Indian society—and an example of a shared business/societal purpose, or "shared purpose." Unilever claims that, by 2015, the Lifebuoy brand had changed the hygiene behavior of tens of millions of consumers across Asia, Africa, and Latin America. The long-term mission is to reach a billion people. Unilever took the Lifebuoy case as a model for its Sustainable Living Plan,[4] which was launched in 2011 in an effort to build its entire business on serving societal and sustainability goals. In other words, Unilever became a shared purpose-driven firm.

5.5 Ben & Jerry's green ice cream

I encountered further example of shared purpose in another Unilever company, Ben & Jerry's, which it had just acquired from the founders Ben Cohen and Jerry Greenfield. One of my SiB partners, Terry Molner, had advised Ben & Jerry's on its sale to Unilever, with a focus on ensuring that Unilever would retain the company's commitment to a societal purpose and the environment. Through a unique acquisition agreement, an independent board of directors was created to provide leadership focused on preserving and expanding Ben & Jerry's social mission, brand integrity, and product quality.

When I met Ben and Jerry on one of our seminars, they told me that they founded their company on the premise of making "great ice cream from great cows growing on great pastures." Bored with their lives and wanting to

Unilever (2012). *Unilever Sustainable Living Plan: Progress Report 2012*. Retrieved from https://www.unilever.com/Images/uslp-progress-report-2012 -fi_tcm13-387367_tcm244-409862_en.pdf.

do something that would be "fun," boyhood friends Cohen and Greenfield decided to start a food business in 1977. At first they considered making bagels but, when the necessary equipment turned out to cost more than they could afford, they settled for ice cream instead. There was only one problem—neither of them knew anything about the business. So they signed up for a US$5 correspondence course in ice cream making offered by the Pennsylvania State University. In May 1978, using US$8,000 of their own money and US$4,000 they'd borrowed, Cohen and Greenfield opened their first Ben & Jerry's Homemade Inc. ice cream scoop shop in a renovated gas station in Burlington, Vermont.

Ben & Jerry's has come a long way from a renovated gas station. With annual sales topping US$200 million, the company reigns as the world's second-largest producer of premium ice cream. But what makes Ben and Jerry so unique is that they redefined corporate philanthropy and CSR. While many companies only go as far as to set aside a portion of their profits for charity, Cohen and Greenfield have actually created products that have in turn created jobs in economically depressed regions, both in the U.S.A. and overseas.

In their quest to initiate innovative ways to improve the quality of life, they have launched flavors such as Chocolate Fudge Brownie, which contains brownies made by homeless and unemployed workers in Yonkers, New York; Wild Maine Blueberry, made with blueberries harvested by Passamaquoddy Indians; and Rainforest Crunch, for which the company buys brazil nuts collected in the Amazon rainforest by indigenous peoples, thereby providing an economically viable alternative to deforestation. In addition, 60% of the profits from that flavor go to environmental groups dedicated to preserving the Amazon rainforest.

Unilever, already the owner of other ice cream brands such as Wall's, was equally capable of making great ice cream but did not have such a strong purpose-based brand. Through its acquisition in 2000 it added Ben & Jerry's to its portfolio to capture the growing market for social brands. As an interesting side-effect, it has inspired generations of Unilever managers how to marry societal purpose with business, and now Unilever is globally recognized as the most sustainable company.[5]

5   For four years in a row until this report (in 2014), Unilever was regarded as the number one corporate sustainability leader, with expert respondents identifying the company as a "leader in integrating sustainability into its business strategy". See http://www.globescan.com/news-and-analysis/press-releases/press-releases-2014/310-global-ex.perts-rank-unilever-number-one-for-sustainability-leadership-in-new-survey.html.

5.6 Turning misery into markets: Medtronic and DSM

There were more examples that gave me hope that capitalism could transform itself into a force for good. Bill George, the former CEO of Medtronic and now a professor at Harvard Business School,[6] told me that as CEO he had taken a similar purpose-based approach to business with considerable success. Together with my SiB partner, Anders Ferguson, I went to see Bill in Lausanne, where he was teaching at Business School IMD, to ask him to participate in the dialogue with the Dalai Lama in 2004. In a coffee shop overlooking Lake Geneva, he told us how he had embedded a sense of purpose at Medtronic, a multinational firm in medical devices such as pacemakers. In a sense, he helped the company to develop a "shared purpose": the company's mission is to restore people to full life and health, while at the same time produce superior medical devices. The way they measured themselves was not only by earnings per share but also by how many people they helped. Under Bill George's leadership, the firm went from 3 million people per year to 10 million people per year who were being restored to a fuller, more active life through the work of Medtronic. The key to success, Bill told me, was to always try to convey this meaning to people in the company because that's what inspires them, not the stock price or the earnings. When the company's mission inspires people, they will produce better-quality work, which in turn will drive financial results.

I found another example close to my home in Royal DSM, a Dutch multinational active in the fields of life science and nutrition. At the end of 2015 DSM employed 20,000 people in 50 countries and posted net sales of €7.7 billion. On top of strong and consistent profits over many years, DSM has been among the leaders in the annual Dow Jones Sustainability Index.[7]

The global sales director, Frederika Tielenius Kruythoff, explained to me what drives DSM's success. The company was founded in 1902 as the Dutch State Mines (hence the acronym DSM) but, when the last coal mines closed in the 1970s, the firm embarked on a process of transformation that continues today. It first diversified into the (petro)chemicals field, and later into essential nutrients such as synthetic vitamins and other ingredients

6   George, B. (2003). *Authentic Leadership: Rediscovering the Secrets of Creating Lasting Value.* San Francisco, CA: Jossey-Bass.

7   Lane, J. (2015, September 13). DSM, KLM rank high in annual Dow Jones Sustainability Index. *Biofuels Digest.* Retrieved from http://www. biofuelsdigest.com/bdigest/2015/09/13/dsm-klm-rank-high-in-annual-dow -jones-sustainability-index/.

for the feed, food, and pharmaceutical industries. The most recent wave of transformation was initiated by Fijke Sijbesma, DSM's CEO, after witnessing poverty firsthand on a trip to Africa. He saw that there is plentiful evidence that improving nutrition in developing countries is fundamental to breaking the cycle of poverty. More specifically, the key lies in providing the right nutrients to pregnant women and their infants. Optimizing the quality of nutrition during a critical thousand-day window of opportunity from conception until a child reaches two years of age has a dramatic impact on its physical and cognitive development, and substantially improves its prospects in adulthood. High-quality nutrition in this phase lays the foundations for a future in which children grow up capable of leading progress in their own communities and countries.

Impressed by both the needs of masses of undernourished people and the market opportunity that this provided, DSM positioned itself as leader in combating malnutrition. Sijbesma said: "As the world's leading producer of micronutrients including vitamins, DSM is taking its responsibility to help solve the world's greatest solvable problem: malnutrition, affecting 2 billion people across the globe."[8] Sijbesma believed that investing in nutrition can not only break the cycle of poverty and build thriving societies and markets, but can also benefit the business objectives of DSM by developing a new market. By establishing partnerships with UN's World Food Programme, DSM turned a global problem into a business growth opportunity of enormous scale. As a business, DSM committed itself to achieving a very tangible goal: to reach 50 million beneficiaries (pregnant and lactating woman and children under two) by 2030.

To show that these objectives are not merely window dressing to boost DSM's sustainability profile, the firm's management board linked its remuneration and executive bonus to DSM's social/environmental performance.[9] These bold steps had an interesting side-effect, Kruythoff said to me: "Our real commitment to societal goals is inspiring our own people. Since taking on fighting malnutrition as part of our mission, our employee engagement has grown substantially. It encourages people to bring their whole selves into their job."

8   DSM (2013, June 8). DSM to contribute to new 2020 global nutrition target. Retrieved from http://www.dsm.com/markets/paint/en_US/news-events/ 2013/06/14-13-dsm-to-contribute-to-new-2020-global-nutrition-target.html.
9   Kolk, A., & Perego, P. (2014). Sustainable bonuses: sign of corporate responsibility or window dressing? *Journal of Business Ethics*, 119(1), 1-15.

5.7 The hidden driver of success

What have these examples in common? On one level, these approaches represent common sense. Businesses are made up of human beings, and like human beings they don't exist for money alone. Humans and business should have some sort of societal benefit in order to flourish. The Dalai Lama says:

> Companies are living, complex organisms and not profit machines. The profit should therefore not be the object of a company, but rather a result of good work. Just like a person can't survive for long without food and water, a company can't survive without profits. But just as we cannot reduce the purpose of a human to eating and drinking alone, we cannot regard companies solely as money-making entities.[10]

This was echoed in the management literature. Jim Collins, in his 2001 bestseller *Good to Great*,[11] explained what distinguishes a great company from a good one:

> Great companies don't exist merely to deliver return to shareholders. Indeed, in a truly great company, profits and cash become like blood and water to a healthy body; they are absolutely essential for life, but they are not the very point of life.

Enduring companies are driven by more than financial profits. This drive can be described as core values, ideology, purpose, mission, or vision—it does not seem to matter what it is called. "The point is not what core purpose you have, but that you have a core purpose at all, and that you build this explicitly into the organization," according to Collins.

The management researchers Rajendra Sisodia, David Wolfe, and Jagdish Sheth write in their 2007 bestseller, *Firms of Endearment*:

> Today's greatest companies are fueled by passion and purpose, not cash. They earn large profits by helping all their stakeholders thrive: customers, investors, employees, partners, communities, and society. These rare, authentic firms of endearment act in powerfully positive ways that stakeholders recognize, value, admire, and even love.[12]

10   Eigendorf, J. (2009, July 16). Dalai Lama—"I am a supporter of globalization". *Die Welt.* Retrieved from http://www.welt.de/english-news/article4133061/ Dalai-Lama-I-am-a-supporter-of-globalization.html.

11   Collins, J. (2001). *Good to Great: Why Some Companies Make the Leap. And Others Don't.* New York, NY: HarperBusiness.

12   Sisodia, R.S., Wolfe, D.B., & Sheth, J.N. (2007). *Firms of Endearment: How World-Class Companies Profit from Passion and Purpose.* New York, NY: Prentice Hall.

What these companies demonstrate is that the purpose of creating value for society is not merely an expression of CSR and philanthropy, of being a "good corporate citizen," but that serving society is at the heart of the business. It is the very reason why these companies are successful, enduring, and great. In other words, when companies focus on creating value for all stakeholders (that is, beyond merely shareholders, to include employees, suppliers, customers, nature, and society) they perform better in financial terms, especially in the long run.

These ideas, in my mind, shattered the mainstream business thinking that equates purpose with profit, the dominant view that is reflected in Milton Friedman's infamous motto: "The business of business is business."[13] I could now see that this thinking is a distortion of how business actually creates value, leading to a dangerous pathway of eroding societal *and* business value. In fact, as I discovered, the hidden driver of long-term business success is a "shared purpose" between business and society. Shared purpose reflects the reality that business and society are intrinsically connected and that it is only this connection that can serve as a sustainable basis for value creation in business.[14]

It also became clear to me that shared purpose starts with leadership. Purposeful leaders drove all the examples of purpose-driven firms. Jim Collins called these "level 5 leaders": "Level 5 leaders are ambitious for the company and what it stands for; they have a sense of purpose beyond their own success."[15] I started to envision a pathway of helping leaders to awaken their organization's shared purpose with society. I sensed leadership development was the best avenue for transforming capitalism and achieving sustainable development. When other executive coaching and training clients came my way, I started to see my work on leadership from a new perspective.

5.8 The purpose gap

But change is hard: there was plenty of resistance. Most companies were not in the league of Unilever or Medtronic. The belief that the "business of

13   Friedman, M. (1970, September 13). The social responsibility of business is to increase its profits. *New York Times*, p. SM17.
14   Hollensbe, E., Wookey, C., Hickey, L., & George, G. (2014). Organizations with purpose. *Academy of Management Journal*, 57(5), 1227-1234.
15   Collins, J. (2001). *Good to Great: Why Some Companies Make the Leap. And Others Don't*. New York, NY: HarperBusiness.

business is business" still reigned supreme in people's minds. For instance, most bankers that I coached, like Rick, did not subscribe to my belief that organizations can be transformed into a force for societal good. They see the needs of their business, and in some cases the needs of their clients, but any constituency beyond that is typically considered an "externality." Traditional managers are trained in the delivery of traditional financial performance and many don't look much further into the future than the end of the fiscal year. Some do not even look beyond the next quarter, which is their next performance target. Societal needs simply do not pop up on these managers' radar.

Among the hundreds of managers that I coached or trained, I learned that there are countless Ricks—all part of the cadre of managers who run the corporate world. Many of them, in spite of good intentions and fancy titles on their business cards, feel trapped in a rat race of performing for the bottom line. When I asked them about the purpose of their work, they often admitted that they had no clear idea of purpose—or that they had it once but lost it—some sort of "purpose gap." Many felt bad because they started out differently—some sort of dream or ideal had guided their first career choice. But now they felt constricted by a strait jacket of achieving key performance indicators that had little to do with why they started their career. The result of this "purpose gap" was a sense of disempowerment.

I discovered that even in the minds of top executives, the ones who you thought would have the final say, there is always someone—or something—else that could be blamed for the purpose gap. It is the fault of the shareholders, the market, the clients, the performance incentives, or—in other words—the "system." As a result, even though these top executives are happy by making good money and acquiring new skills, many tend to feel disempowered within modern capitalism.

Obviously, this situation is not conducive to changing the corporate culture toward greater sustainability. I designed a leadership development program for a large sportswear company and placed a discussion about sustainability on the agenda, but the HR director asked me to remove it because sustainability was not in the functional responsibility of any of the program participants. The leaders that I had to train just needed better leadership skills, and discussions on sustainability would distract from that, he said.

Many managers react skeptically when asked about implementing sustainability goals. The key challenge, they say, is that the incentives are simply not there yet. There is greater awareness of social and environmental impacts at the leadership level, along with improved communication about the company's approach to managing these impacts. However, in many companies, much of this activity is superficial and largely in the interests of managing the reputation (i.e., "we can talk all day about our socially responsible actions as long as it doesn't affect our business model or the bottom line"). While some shareholders may tacitly recognize some "value" in a sustainable company, the investment industry has not given strong value to social and environmental impacts in the market. For many, therefore, sustainability is not considered a force for transformation. The "system" simply does not permit it.

I felt it was ironic that I had encountered the same sort of disempowerment among people at the lowest economic levels of society in developing countries. A coffee-bean planter in rural Indonesia talked about the "system" (often embodied as the middleman or the money lenders) in similar terms as the senior manager in a London bank. It was clear that the system of global capitalism had fostered a culture of disempowerment at both the top and the bottom of the pyramid. Nobody within the system feels as though they are in charge. The Tibetan teacher Sogyal Rimpoche, speaking to a group of business leaders, put it in these words: "You are very successful managers, with busy and exciting lives, but the question is this—is there somebody at home? Do you have a sense of who you are beyond your busy life?"[16] We had created this powerful machine of production and consumption, run by a class of clever managers, but who was in charge? And who was in charge of these managers' minds?

The metaphor "nobody at home" struck a chord. I recalled how powerless I felt when I had lost my purpose at the bank. The only thing that had mattered at that point was survival, the next deal, the next report to my boss, the next salary review. It was clear that if I worked with executives that were only interested in meeting the expectations of their boss, or saving their marriage, I would miss my goal.

How to change this? How to transform this vicious cycle of disempowerment, of blaming the system, and of being "not at home"? How could we bridge the purpose gap and instill or awaken a deeper purpose in business leaders? How can we expand the limited sense of purpose to include the needs of a wider range of stakeholders and make it a "shared purpose"? How could we help managers to see that this purpose-based approach was not a luxury but a necessity in the face their increasing dependence on external stakeholders?

16 Sogyal, R. (2002). *The Tibetan Book of Living and Dying*. New York, NY: HarperCollins.

5.9 The science of transformation

In order to answer these questions, I deepened my research on organizational change. How could I help people to move from a "small" purpose that merely focused on oneself to a "big" and shared purpose, focused on the interests of self and others and even on benefiting society at large?

I was most intrigued by how the revolutionary insights in neuroscience (as we discussed in Chapter 3), could be applied to business, as it was the language of "hard" science that resonated in the secular business world. I found the phenomenon of neuroplasticity most exciting. Neuroplasticity quite literally means that brains are "plastic," i.e., they can be changed through practices such as meditation and reflection. In other words, negative patterns in the brain (such as selfishness and short-term thinking) could be transformed into positive patterns (such as altruism and long-term thinking). If this applies to human brains, it should logically follow that plasticity must also be true for organizations because they are made up of human beings "with brains." If beings can be transformed, so should organizations and economic systems. Said differently, organizations and economic systems are social constructs. If humans can be changed, social constructs can be changed. It is just a matter of aligning the ideas and practices of the people involved.

It seemed a fortuitous coincidence that both science and business started to speak in terms of the transformation of people. I found this was a very hopeful perspective. At the Irvine dialogue, David Cooperrider had called for a positive revolution in leadership development—I realized that was my call too. In addition to my interest in the field of emotional intelligence and positive psychology at the Teleos Leadership Institute, I studied with several other innovative leadership experts in their respective fields. Richard Barrett, author of *Liberating the Corporate Soul*,[17] invited me to join his think-tank on whole systems change with management thinkers such as Don Beck,[18]

---

17   Barrett, R. (1998). *Liberating the Corporate Soul: Building a Visionary Organization*. London: Routledge. The outcomes of the think-tank on whole systems change have been laid down in: Barrett, R. (2009). *Building a Values-Driven Organization: A Whole-Systems Approach to Culture Transformation*. London: Butterworth-Heinemann; see also https://www.valuescentre.com.

18   Beck, D.E., & Cowan, C.C. (1996). *Spiral Dynamics: Mastering Values, Leadership, and Change*. Cambridge, MA: Blackwell.

Michael Rennie,[19] and Frank Dixon.[20] I also studied with the leadership experts Peter Senge,[21] Otto Sharmer,[22] Miki Walleczek, Sue Cheshire,[23] and Erica Ariel Fox.[24]

While I learned immensely from their pioneering work, I was struck by the realization that certain essential parts were still missing from these new schools of practice. The key problem, I felt, was that they could not explain the connection between inner and outer transformation. In my own consulting practice I had observed that, when leaders transform, it was not guaranteed that they would also transform their companies. Some leaders, after experiencing a sense of purpose in a leadership development program, preferred to resign from their firms. Rather than bringing their purpose into business, they renounced their role as a leader to work on inner transformation. Apparently, the purpose gap with their companies was too big. Conversely, when managers set out to transform their organizations, they didn't necessarily turn into better human beings. Many remain stuck in the middle, lacking the tools to match inner intentions with outer conditions.

> The key problem was that they could not explain the connection between inner and outer transformation.

My SiB friend Anders Ferguson introduced me to someone who had identified the same gap. Dr. Joe Loizzo, a Harvard-trained psychiatrist, Buddhist scholar, and founder of the Nalanda Center for Contemplative Science,[25] and whose work on integrating Buddhism with modern psychology and therapy I admired, not only agreed with my observations but also pointed to Buddhist theory and practice as the way to overcome this. In a sense, the insight that I

19   See http://www.mckinsey.com/global-locations/europe-and-middleeast/middle-east/en/our-people/michael-rennie.

20   Dixon, F. (forthcoming). *Global System Change: Achieving Sustainability and Real Prosperity*. In press. See also http://www.globalsystemchange.com.

21   Senge, P.M. (2008). *The Necessary Revolution: How Individuals and Organizations Are Working Together to Create a Sustainable World*. New York, NY: Broadway Books. Peter founded the Society for Organizational Learning: http://www.solonline.org.

22   Sharmer, C.O. (2008). *Theory U: Leading from the Future as it Emerges*. San Francisco, CA: Berrett-Koehler Publishers.

23   See http://globalleadersacademy.com.

24   Fox, E.A. (2013). *Winning From Within: Breakthrough Method for Leading, Living, and Lasting Change*. New York, NY: HarperCollins.

25   Loizzo, J. (2012). *Sustainable Happiness: The Mind Science of Well-being, Inspiration and Compassion*. New York, NY: Routledge. See also the Nalanda Center for Contemplative Science: http://www.nalandainstitute.org.

gained from Joe was a simple and fundamental Buddhist truth: since the nature of reality is interconnectedness, people can only be effective to the extent that they understand and operate according to the reality of interconnectedness. It followed that, for any leadership model to be relevant today, it needs to be rooted in the principle of interconnectedness, which underlies both personal and organizational transformation. This, in fact, was the key lesson I had been learning from the Dalai Lama, but now I could see how it could be applied to the secular world of leadership transformation in business.

This insight is best expressed in two fundamental Buddhist principles: wisdom and compassion. In the Buddhist view, wisdom is the correct understanding of how the phenomenological world exists and operates, namely as an interconnected system. Because of the interconnected nature of reality, there cannot be a separate thing called the "self" that exists independently from others and from nature. It is this illusory sense of self on which people generally place all their hopes and fears, which causes them to revolve in an endless cycle of suffering, known in Buddhism as *samsara*. Wisdom, therefore, is concerned with discovering the true interconnected and "selfless" nature of all phenomena. Likewise, compassion is the best strategy to deal with this selfless interconnected reality. It is these insights that will liberate people from *samsara* into *nirwana*.[26]

5.10 Bodhisattva leadership

In Buddhism, someone who decides to take on the path of cultivating his mind to the fullest is called a Bodhisattva.[27] Given the interconnected nature of reality, the Bodhisattva does not make a distinction between self and others. While the principal focus is the development of his own mind, he is equally concerned with the world around him. Given that we and others are interconnected and interdependent, and that our happiness in the ultimate sense relies on the happiness of others, compassion is considered a natural state. The Bodhisattva, therefore, works on his mind to develop wisdom while practicing compassion.

26   Dalai Lama (2002). *The Meaning of Life from a Buddhist Perspective*. Boston, MA: Wisdom Publications.
27   Wallace, B.A. (1993). *Buddhism from the Ground Up*. Boston, MA: Wisdom Publications.

The Bodhisattva—which literally means a "being on the path of awakening"—can be described as a spiritual warrior: he essentially is an "inner warrior," someone fighting the *inner* enemies of ignorance, selfishness, greed, and anger, rather than fighting any *outer* enemies.[28] Yet his purpose is serving the interconnected world at large. The Bodhisattva is the Buddhist metaphor for leadership, embracing both inner and outer leadership. In Asian history, therefore, rather than merely renouncing the outer world, Bodhisattva warriors often intentionally took on roles in the outer world as statesmen, teachers, artists, and writers, which allowed them to serve others while simultaneously engaging in an inner practice to transform their own mind.[29] The Bodhisattva is the ideal of the Buddhist tradition that the Dalai Lama belongs to[30]—in fact, in my mind, the present Dalai Lama represents and embodies this tradition to the fullest.

Since the Bodhisattva, unlike any other leadership archetype, is based on the wisdom that understands the interconnected nature of reality, I felt that the concept could serve as inspiration for leadership that has to deal with complex interconnected challenges such as sustainability, for the principles behind the Bodhisattva practice are similar to those driving sustainability. I therefore felt that the Bodhisattva could be a role model for the new type of leadership that we need in business: a modern-day warrior for sustainable system change at three interrelated levels: the individual, the organization, and the collective, spanning both the "inner" and the "outer" dimensions.

How can we translate these ideas to the modern context of business? If a company's leadership is oblivious to the needs of its stakeholders, on whom it depends for its long-term survival, the firm is bound to experience trouble at some point in time. For example, customers will stop trusting the brand. In contrast, if the leadership of the company understands and caters for the needs of its stakeholders (in balance, of course), it is likely to be successful in the longer term, simply because this mind-set allows them to see more and better adapt to future needs. Wisdom is the firm's ability to see its interdependence with other stakeholders, while compassion is the ability to serve stakeholders' needs. Since the leadership of companies such as Unilever

28 Thurman, R. (1997). *Inner Revolution: Life, Liberty and the Pursuit of Real Happiness*. New York, NY: Riverhead Books; Trungpa, C. (1984). *Shambhala: The Sacred Path of the Warrior*. Boulder, CO: Shambhala Publications.

29 Loizzo, J. (2006). Renewing the Nalanda legacy: science, religion and objectivity in Buddhism and the West. *Religion East & West*, 6, 101-121.

30 Mullin, G.H. (2001). *The Fourteen Dalai Lamas: A Sacred Legacy of Reincarnation*. Santa Fe, NM: Clear Light Publishers.

and Medtronic employ such "compassionate" stakeholder-oriented mind-sets, they are likely to be more successful (in the long run), than those who employ more limited, self-oriented mind-sets.

In this way, Buddhist principles of wisdom and compassion can make sense to business strategy. I started to envision the possibility of training leaders to become "sustainability warriors." While much of the current global sustainability crisis is unprecedented, it appears that the intercon-nected and interdependent worldview that solutions to the crisis are call-ing for has been recognized and practiced by leaders in Buddhist-inspired civilizations such as Tibet. The preservation of the contemplative traditions like the Bodhisattva path suggests that the leadership practice we need to employ in order to deal with this new paradigm may already exist.

The only problem, I thought, is that the terminology used to express these ideas is cloaked in religious connotations. This does not work for the secular business mind. Expert translation between the two worlds is needed. It then dawned on me that perhaps I, by having a foot in both worlds, had a role to play in designing a theory and practice of leadership suitable for the system change that is needed, and aligning it with proven concepts and methods from Buddhism.

I decided to experiment. One year, sitting on a mountain slope in front of a Buddhist monastery in Tibet, I conceived of a leadership journey to Tibet. The next year, in the summer of 2007, I traveled with a dozen busi-nesspeople to Eastern Tibet to engage in a two-week program of leadership development using a combination of practices from Buddhism and the new schools of leadership transformation. In the spirit of the Bodhisattva, the inner transformation process was designed for the sake of outer transforma-tion. Inner and outer worked two ways: while traveling in the inspirational landscape, we asked people to reflect on their internal mindscape—the frames of references through which they look at the world—and how these inspirational outer experiences can bring inspiration to their lives and jobs. Meeting with Tibetan Buddhist masters and nomadic farmers, and sitting silently gazing at vast skies or nightly campfires, several participants were able to create a new vision for their work and life. In fact, all participants enjoyed some sort of mind-shifting experience. Rein Heddema, a former colleague and human resource director at ABN AMRO Bank who joined me on the first journey, wrote to me some years afterwards: "This journey cre-ated a profound positive change for me in my work and private life. Even many years later, I still experience the beneficial effects. It was a true turn-around, a once-in-a-lifetime experience."

I repeated this experiment every year. Later I added other destinations, including Bhutan, Mongolia, and, closer to home, in the Swiss Alps. In Bhutan, we guided the participants in an exploration of the concept of gross national happiness. The culture and pristine landscape of Bhutan provided inspiration for participants to redesign their own life and work into a more positive and joyful direction. How could they as leaders, while learning from Bhutan's unique development philosophy, transform their organizations into a force for societal happiness? Designing and leading these journeys was a very satisfying experience for me. Finally, I had found a professional expression of the many threads that made up my work—sustainability, leadership, business, and Buddhism, fused into one program.

5.11 Global meltdown

Little did I know that we were approaching the end of an era. On all work fronts, cracks appeared. In the run-up to the Beijing Olympics in 2008, when the Olympic torch went from country to country, Tibetans staged well-publicized protests, which were highlighted in the media all over the world. Suddenly, Beijing, which had wanted to use the Olympic Games to brush up its reputation as an ultra-modern capital of an ancient civilization, had been tarnished by the Tibet issue. When the games were over, Beijing used its full force to crack down on Tibet. China's leadership perceived the Tibetan protests as masterminded by the Dalai Lama. Tibetans with any ties to foreigners or with any foreign funding were removed from their jobs. Some ended up in jail.

Foreign visitors were no longer welcome, especially those with connections to the Dalai Lama. I had to learn this the hard way on what was to be my last trip to Tibet. On the night after arrival in Lhasa, the Chinese security police knocked on my hotel room door and summoned me to immediately follow them to their office. They took my passport and told me that they had a file on me. The file stated, they said, that I was a confident of the Dalai Lama. I tried to remain calm and friendly and used my Chinese language to connect with the police officers on a personal level, some of whom were Tibetan. It may have worked. They released me after a few hours with the warning that I should stick to a very strict itinerary and keep my official guide with me all the time. I was lucky. But at the same time I realized that, with the new regime in place, I had little chance of continuing the work that we had started a decade earlier across the Tibetan plateau: it would jeopardize the safety of the Tibetans who worked with us.

But a larger crack was showing. The speculative boom in the U.S. mortgage-backed security market had begun to slide and, in September 2008, the U.S. investment banks Lehman Brothers and Washington Mutual went bankrupt. With asset values of US$691 billion and US$327.9 billion respectively, they became the world's largest and second-largest bankruptcies in history.[31] This led to the near collapse of the financial sector in the U.S.A. and Europe in the fall of 2008, when it turned out that most globally operating banks were not only overly exposed to the mortgage market but also to each other—as one interconnected system. The European and U.S. governments stood behind many more troubled firms and offered billions of dollars to prevent a total economic collapse.

However, by 2009, the crisis had deepened and spread to governments, especially the European countries of Greece, Ireland, and Portugal. They too were overly indebted. The situation was dubbed the "debt crisis" and later, when the creditworthiness of the European monetary union was called into question, the term "Euro-crisis" emerged. The crisis reminded me of the Asian financial crisis a decade earlier, as it was characterized by identical patterns—overexposure through speculation, sudden eruption of distrust, and outsourcing costs to taxpayers through bailouts. But this time the scale was global. By 2009, it was unclear where this would all lead, but many observers suggested that this was the end of financial capitalism as we know it. It would be the beginning of a new economic paradigm, but nobody was sure what that would mean.

The president of the U.S. Federal Reserve, Alan Greenspan, who received much of the blame for keeping interest rates low and allowing banks to take irresponsible risks, was replaced by Ben Bernanke. When he took office, Bernanke was in a reflective mood: what is the future of capitalism? In a speech he suggested that Western economists should learn from the Buddhist country of Bhutan, which had developed an alternative economic model called gross national happiness.[32] I found that ironic. While Bhutan was trying to learn from the West, the leading Western economy—the leader of capitalism—was now turning to the small isolated Himalayan kingdom

31   Fortune (2009). The 10 largest U.S. bankruptcies. *Fortune*. Retrieved from http://archive.fortune.com/galleries/2009/fortune/0905/gallery.largest_bankruptcies.fortune/.

32   Bernanke, B. (2010). The economics of happiness. University of South Carolina commencement ceremony, Columbia, SC. May 8, 2010. Retrieved from http://www.federalreserve.gov/newsevents/speech/bernanke20120806a.htm.

of Bhutan for guidance. Buddhism and banking were finally finding each other, it seemed. In 2012, the General Assembly of the UN even accepted a resolution by Bhutan to make the conscious pursuit of happiness a fundamental human goal, in the resolution "Happiness: towards a holistic approach to development."[33]

It appeared to me that the ideas we had been exploring with the Dalai Lama were starting to have an impact on the thinking of policy-makers in both the East and West and that the world was finally—partly prompted by the financial crash—waking up to the need to create alternative economic and business models.

The first decade of the millennium also witnessed the worsening of the environmental crisis. Thanks to the movie, *An Inconvenient Truth*, climate change was no longer a taboo topic. Many corporations recognized the challenge of sustainability and appointed chief sustainability officers in their headquarters. By the summer of 2009, many people were still hopeful that the ambitious UN Climate Change Summit in Copenhagen, due at the end of that year, would bring the world closer to tackling the problem. However, fears that the unfolding economic crisis would undermine this intention started to grow.

It was clearly time to reconnect with the Dalai Lama and to hear what he had to say about the crisis and the way forward for humanity. Given the development in my own thinking, I suggested that the focus of the next dialogue should be on leadership for creating sustainable economic systems and organizations; hence we titled the next conference with the Dalai Lama "Leadership for A Sustainable World."

33   United Nations (2012, July 12). Resolution 66/281 on Happiness. New York, NY: UN General Assembly.

The author welcoming the Dalai Lama when he arrived at the dialogue in The Hague, 2009.

6
Third dialogue: Leadership for a Sustainable World (The Hague, 2009)

The third public dialogue with the Dalai Lama had a somewhat different format. In order to provide a voice for many more people and perspectives, the dialogue became part of a whole-day conference that explored the theme of leadership in creating a sustainable world. Questions were generated through a participatory process and these were later presented to the Dalai Lama. The key themes that emerged from the discussions help us to see the role of business in a new light:

- The eruption of the crisis, in essence, was a result of using outdated narrow concepts to deal with a new complex interdependent world. In the 21st century, new conceptual models are needed that go beyond the ideas of endless economic growth and profit at any cost, and the accompanying attitude of "us against them."

- The crisis has shown the importance of developing a positive compassionate mind-set. This more altruistic mind-set will also help individuals be more resilient and happy, and less dependent on external factors that we believe make us happy, such as money.

- Science has confirmed that love and compassion have a profound effect on personal health. For that reason, there is a need for secular ethics that can be applied to general education and business.

- Everybody has a role to play and the potential to make a difference. Economic and social systems can only be changed by people. Global companies are in an ideal position to take an active role. Ultimately, solutions will come from close cooperation and collaboration of many people and organizations.

The date that the Dalai Lama had given us for a next meeting—June 5, 2009—coincided with the World Environment Day, which we considered to be an auspicious sign.

Among our 500 guests were several people who had been present at the first dialogue with the Dalai Lama in Amsterdam ten years earlier. One of them was Ruud Lubbers, former prime minister of the Netherlands, who opened the conversation.

Ruud Lubbers: Ten years ago the Dalai Lama came to the Netherlands and addressed us at the symposium, "Compassion or Competition." Today he is again in our country to inspire us, leaders of business and society, on how we can develop leadership for a sustainable world. A precious day!

How can we contribute to a sustainable world? It is already a long time ago since Albert Einstein said that the problems that exist in the world today cannot be solved by the level of thinking that created them. In 1988, the Brundtland commission, in the publication *Our Common Future*,[1] gave voice to a level of understanding that recognized sustainability as a key challenge for humankind, focusing on intergenerational solidarity and responsibility.

After the end of the Cold War, market economies all over the world over-ignited economic growth in a rapidly globalizing world. But by 1992, at the United Nations Conference on Environment and Development in Rio de Janeiro, civil society took a firm stand and proclaimed the "Earth Charter."[2] As a result, the Rio de Janeiro conference became known as the Earth Summit. In the words of the Earth Charter: "We are one human family and one earth community with a common destiny."

1   World Commission on Environment and Development (1987). *Our Common Future*. New York, NY: United Nations.
2   See http://www.earthcharter.org.

It took more than seven years of worldwide consultations with civil society groups and leaders, including with religious leaders from the Baha'i, Buddhist, Christian, Confucian, Hindu, Indigenous, Islamic, Jain, and Shinto traditions, to finally complete the Earth Charter document in 2000. Just one year earlier, at the symposium "Compassion or Competition," the Dalai Lama spoke about the importance of compassion in globalizing business, and Hazel Henderson spoke about the dangers of "casino capitalism."

Now, ten years later, we see that indeed an excessive greed-oriented financial world resulted in a major crisis and this in turn lead to the collapse of the financial system and the subsequent recession in OECD countries. This is especially tragic because in the last decade globalization brought about something positive, namely corporate social responsibility. More and more transnational companies, often under pressure from civil society and the individuals working in those very companies, shifted their mission statements and business practices toward becoming more responsible and sustainable. And important initiatives such as the UN Global Compact and the Global Reporting Initiative emerged.

The Social Economic Council of the Netherlands has indicated that the time has come to add a fourth P to the three common Ps of people, planet, and profit, namely the P of *pneuma*. The Greek word *pneuma* is reflected in the Latin root of the word "spiritual," which means "breath." In Eastern traditions, harmony can be restored through meditation and this includes an awareness of your breath. I therefore rejoice that an outstanding spiritual leader from the East, the Dalai Lama, will assist us in understanding that, in order for us to develop humanity in the secular world, we need to practice spirituality as well.

> In order for us to develop humanity in the secular world, we need to practice spirituality as well.

This conference takes place at a moment when the world badly needs to recover from the economic recession. But let me be clear. I believe that this can only happen if leaders—"we the people" in business, religion, and spirituality—go for a new "green deal." If we want to move from individual rights to truly taking collective responsibility for a sustainable world, we indeed need to add the Earth Charter to the Universal Declaration of Human Rights. I want to conclude with the words of the late Earth Charter commissioner, Kamla Chowdhry: "In the end, the world will be saved not by our wars and our military leaders, not by our science or technology, not by our industrial magnates, not even by our politicians, but by our saints and spiritual leaders, who lead with integrity."

6.1 Learning from the financial crisis

Sander Tideman: It's a great privilege and a great joy to once more have the opportunity to discuss with Your Holiness the topic of leadership for business, economics, and society, particularly with the aim of creating a sustainable world. Since the financial crisis erupted nearly a year ago, it has become clear that this topic is relevant to all of us simply because the world is becoming increasingly interconnected and interdependent. We are all feeling the effects of the crisis, which from our perspective had started far away in the subprime mortgage market in the U.S.A. The causes may seem obscure, but the consequences are very clear. These economic and financial problems, and also ecological problems such as climate change, water shortages, and biodiversity loss, now require us to take up a larger responsibility than before.

It is clear that these problems will not simply disappear, even if we get the economy back in order. After all, we can't economically "grow" ourselves out of global warming because pursuing economic growth has caused it. The model of economic growth that we have quite blindly pursued over the last few decades now requires updating, yet we don't have any credible alternative model.

We have discussed these issues with you in earlier meetings, so in that sense there is nothing new. But the fact is that the problems have become more complex and widespread and—as I said—they are now affecting us all. They have become more systemic and intermingled. So we are again looking for guidance from you. The outcome of this morning's discussion is that we can bring all this down to these questions: how can we—leaders in business and society—create a sustainable world? How do we create a more compassionate society and a more sustainable economy?

The Dalai Lama: Since most people here work in the field of the economy, I think I am the wrong person to answer your questions. If I were to take responsibility for the economy, then I fear that within a week or so the economy would collapse! Of course, through earlier dialogues, and through discussions with my friend, Laurens van den Muyzenberg, with whom I co-authored a book on responsibility in business [*The Leader's Way*][3] and who is in the audience today, I learned some things about big corporations and

---

3   Dalai Lama, & van den Muyzenberg, L. (2009). *The Leader's Way: The Art of Making the Right Decisions in Our Careers, Our Companies, and the World at Large.* New York, NY: Crown Business.

big macroeconomic issues. When I worked on this book, I learned that ethics are not only relevant but also very important to economics. Alongside ethics, in business and economics you need transparency and truthfulness. It is very essential if you want to have long-term success.

Now we are experiencing a major financial crisis. Sometimes I feel confused when people speak about market forces; it seems that these forces are something beyond human control. But the market in itself is a human creation. This is something that is created by us human beings, and yet it is somehow out of our control.

Last year, when the financial crisis started, I met one of my longtime friends, an Italian businessman. Out of curiosity, I asked him: "What are the causes of this global financial crisis?" He answered: "The cause of the crisis is greed. Greed first and speculation second." I understand that speculation means that you make a business decision, maybe on the basis of a naïve wish or, worse, on the basis of greed without understanding what is really going on. This means that the motivation behind the business decision is not right. My friend's answer made me realize that, with regard to the larger economic system, these moral things are very important. It is not simply up to market forces which are supposedly beyond our control.

So obviously, this means that all those of you who work in corporate organizations, and all those who are dealing with money, you have to be transparent right from the beginning. You have to make clear to everyone involved when there are difficulties, what is going on, what the risks are and so on. And if things go wrong, then don't pretend that everything is going smoothly. This is a matter of truthfulness, honesty, and transparency.

I believe that every human action ultimately relates to our motivation. Generally, when your motivation is sincere, if you have a compassionate motivation, then even if the work is not very successful, that is OK! Then the outcome doesn't matter. I believe that in most cases, if you work with a sincere compassionate motivation, your mind will be more balanced, more neutral. In that case, you can see the things more objectively. If your mind is full of greed, with too much attachment, or with too much hatred, you can't see reality objectively.

These strong emotions distort reality; they cause a mental projection to happen. I think in the business field—or any human activity for that matter—you should take a realistic approach. Otherwise nothing happens. In order to find a realistic method, you must understand the reality. This is very essential. If you look at a situation from just one angle, you cannot see reality. So try to look at it from three, four, or six angles; look at it from a broader

The Dalai Lama meets conference participants.

long-term perspective and then you can see reality more clearly. Especially today, with so many complex situations, it is very important to develop a more holistic picture. This will give you a more realistic understanding and also a more balanced mind.

When your mind is full of desire, full of hatred, full of a sense of competition, or full of jealousy, it cannot be calm. So that is why compassion is important. Now here I don't mean compassion in the usual sense of love or romance, a feeling in which you expect something in return from the other. This feeling is full of attachment. That is a limited, biased compassion, which prevents you from seeing reality clearly. Genuine compassion is a mental state that is based on that state of limited compassion, but then through further training, using reasoning and common sense, we can strengthen this limited compassion and we can transform it into unlimited, unbiased compassion. Limited compassion is oriented to the other's reaction toward you, to what they give you, while unbiased compassion is oriented toward the genuine well-being of the other, regardless of what they give you.

This type of compassion can even be extended to your enemy. Because your enemy is also a human being, regardless of their attitude toward you, they want happiness and well-being. So that type of compassion, which is oriented toward happiness of others, is really helpful to maintain our peace of mind. Actually, medical science has indicated that this type of compassion is helpful for maintaining good physical health as well.

In short, the practice of compassion brings many benefits. In this context, compassion is helpful in seeing reality. Every human being—including those in business—is driven by the desire to be happy and avoid suffering. This is a fact that everybody can agree on. Therefore, I think that there will be benefit from developing a more compassionate mind in business.

Finally, I would like to share some thoughts on the larger economic system. As I mentioned, I was collaborating with Laurens van den Muyzenberg on the book *The Leader's Way*. Initially, we had some sort of disagreement: I am more attracted to Marxist socialism than to capitalism. So at one time he brought the book by Karl Marx, *Das Kapital*, and he told me to read this. "You will see that Marxism is scientifically wrong," he said. But after reading it, I still feel that between capitalism and Marxist socialism, I prefer socialism. However, I totally oppose Leninism, because it not only advocates equal distribution, but also centralized power as a means to achieve this! This undermined the original Marxist thinking. It has become a totalitarian ideology, which has caused suffering for millions of people.

Someone recently described me as a Marxist monk. In Marxist theory, the focus is the allocation of wealth. From a moral perspective, this is correct. Capitalism, on the other hand, values the accruement of wealth—the allocation of wealth doesn't matter. In a worst-case scenario, the rich will keep getting richer while the poor keep getting poorer. So I believe capitalism needs to be combined with the principles of responsibility and sharing. This is what I call a "responsible free-market economy."

This leads me to conclude that the current financial crisis could teach us to look beyond material values and unrealistic expectations of limitless growth. When things go seriously wrong, it is often because a new reality is still being viewed with outdated concepts, and this is certainly the case with the economy today.

The new reality is one in which the challenges facing humanity are "beyond individual effort" and our interdependences have become even starker. The gaps between our perception and the new reality are based on having concepts from the previous century in our minds, and this wrong perception creates the wrong approach. In

> The current financial crisis could teach us to look beyond material values and unrealistic expectations of limitless growth. When things go seriously wrong, it is often because a new reality is still being viewed with outdated concepts, and this is certainly the case with the economy today.

business, these 20th-century concepts include management approaches that focus on maximizing short-term profits [not on delivering longer-term "triple bottom line" outcomes], ignoring activists or critical stakeholders, only caring about a regulatory license [not the broader "social license"] and an "us and them" approach to problem-solving.

In business terms, a wrong perception of reality is likely to lead to the wrong strategy. It is increasingly likely to lead to a poor reputation and missed opportunities for the organization. Worst of all, it is likely to prevent an effective approach that addresses key global challenges.

> In business terms, a wrong perception of reality is likely to lead to the wrong strategy. It is increasingly likely to lead to a poor reputation and missed opportunities for the organization. Worst of all, it is likely to prevent an effective approach that addresses key global challenges.

6.2 Creating a sustainable economy: where does the leadership come?

Sander Tideman: As you said, the current financial crisis is causing much concern in the West right now, as it reflects a new reality that does not fit into our current, yet outdated, concepts. You emphasized the role that values, or the lack of them (especially overvaluing money and undervaluing other values such as friendship, ethics, transparency, and so on), has played in bringing about the crisis. I think we would all agree to that. But is it only about changing individual people's mind or can governments, on a national and international level, do something? The crisis has taught us also that we cannot leave it to individual market players. What role does the State play in creating a healthy and sustainable economy?

The Dalai Lama: That is a very hard question. Obviously, governments have a responsibility to maintain clear rules and standards of fairness and transparency. However, I have learned from the situation in China, in which the State forced a form of socialism on its people with very mixed results, that not all solutions can be established at a national level. Nations can cause a lot of damage. This is why I warn against expecting too much from governments when it comes to the redistribution or regulation of the financial markets. People will always find ways around rules and laws—even if they are the best rules or laws. Or do you think it was the lack of regulations that lead to this financial crisis? I have been told that the rules in the United States were not that bad; but responsible action demands more than law abidance.

In the end it comes down to every single individual; it is dependent on each individual's sense of moral responsibility, self-discipline, and values. In my assessment, the financial crisis isn't purely a crisis of the market economy, but rather a crisis of values.

> The financial crisis isn't purely a crisis of the market economy, but rather a crisis of values.

Sander Tideman: So you are saying that the best approach is to focus on the development of individuals, for example by developing different types of educational program, such as those that emphasize responsibility, values, morality and ethics, etc.?

The Dalai Lama: Yes, history, math, languages, and economics—these are all subjects for the brain. But responsibility—moral responsibilities, responsibilities regarding society—these are things that come from the heart. This attitude of responsibility, combined with the power of the brain, is what the world needs, including our government and large companies, because it will make them more effective. I will give you an example: we Tibetans believe that our national issue with China can only be resolved nonviolently. This is what we preach from kindergarten and throughout the entire education of an individual. When a Tibetan is confronted with a conflict, his reaction should immediately be: how can I resolve this in dialogue? It is important to us that young people in our schools understand that violence is the wrong way, that violence cannot solve problems. This attitude has become a part of many Tibetans' life through education and training. The same needs to happen as regards the economy and our financial systems.

> This attitude of responsibility, combined with the power of the brain, is what the world needs, including our government and large companies, because it will make them more effective.

Sander Tideman: I understand that this is the best long-term approach, but many of us here are impatient and are looking for some immediate steps to take. Can we speed up the transition to a more "responsible free-market economy"? And many of us work for business organizations that have global networks through which they can impact many others. Is there a special role for them?

The Dalai Lama: First, with regard to speeding up, I think this financial and economic crisis will help us to make this transition faster because those people who only think about money—even dream about it—are affected the most by the crisis. The crisis is terrible for many people, but it also shows that the value of money is limited, and the insecurity inherent in money and financial systems is huge. Inner values such as friendship, trust, honesty, and compassion are much more reliable than money—they always bring happiness and strength, regardless of any outer crises.

> Global corporations are in an ideal position to support developing countries in closing the gap between their economies and those of the most wealthy nations.

Then, when looking for solutions, when we search for organizations that have the capacity and ability to improve our world, global companies are at the top of the list. In particular, global corporations are in an ideal position to support developing countries in closing the gap between their economies and those of the most wealthy nations. Some people believe that global companies should be dismantled because they are the source of all the problems, but I think this view is one-sided. Clearly, we should have better rules to prevent companies from destroying the environment and public goods such as climate, water, food, and biodiversity, or using them solely for their own financial advantage. But we cannot turn the clock backwards. Globalization, including the rise of globalized companies, has occurred because technology has made it possible.

> we should focus on using globalization and global companies to solve global challenges.

Globalization is not a problem in itself because it also has given us many benefits. Now we should focus on using globalization and global companies to solve global challenges.

6.3 Is money leading us?

Peter Blom: This question comes from a banker: what is the true meaning of money? You mentioned that we overvalue the importance of money, so much so that we are too attached to it, which I agree with. But especially now, after the financial crisis, some people see it as evil, as the root of all economic problems, whilst others of course still view it as the essential fuel for the economy. Clearly, the crisis has shown us that our understanding of what money is, and what it can mean for our society, needs to be reviewed. So my question is: how can we make money into a good servant of society, a compassionate servant, rather than this thing that is mastering us and has caused us to create this crisis? In other words, how can we lead money rather than allow money to lead us?

The Dalai Lama: There is nothing intrinsically wrong with money. In Tibet, people speak of money as "a means to accomplish one's aims" rather than the aim itself. Therefore, the problem is not with money but with our attitudes toward money.

In fact, those who are most focused on money are often most unsettled by it too; in the current economic climate it is probably these people who are having the most sleepless nights. This is because they have grown attached to money itself and have lost a balance between money and other values. So my advice to the business leaders in the audience is to work on establishing a balanced approach to money and to also pay close attention to cultivating inner values.

If you are too attached to money, this in itself will create problems. I am sure that those who have had such an attachment have suffered much more from the financial crisis than those who don't have such an attachment. The problem here is not money, but the attachment to it. This applies to other objects as well. For example, I am a Buddhist and if I have too much attachment toward Buddhism, then when I meet with people of different religions, this attachment creates a problem. This is the same with scientists. Some are so attached to their own views or those of their discipline, they are not open to listening to other views.

In today's economy, we need money. Without money, you can't develop your business and economy. In fact, to solve certain social problems such as poverty and sickness, we need a lot of money. So the ability to make money is very important. But if we become too attached, then it will create emotional problems. I know of an American billionaire with a good reputation, but as a person was not happy at all. In other words, wealth and reputation don't deliver happiness.

Attachment causes us to have a mistaken view of reality. We expect the object to which we are attached will give us happiness. In Buddhist texts it is explained that when we face a problem and you know what to do to overcome that problem, then there is no need to worry; the only thing you have to do is to apply effort to solve the problem. But if the problem is such that there is no way to overcome it, then there is no use worrying. Accept it as fact. This is what I call the realistic approach. I think this is the source of a calm mind.

This mental attitude of not being attached can make a very big difference for our mental calmness. It is not the external object itself, for example whether we lose money or not, that makes us happy. In affluent societies we can become attached to so many beautiful things. In those societies I strongly recommend people to pay more attention to their inner attitude, to their mind. Everybody can do that. Everybody can develop a mind that is less attached to external objects. This will make you less susceptible to external problems.

There was a Tibetan monk who spent 18 years in a Chinese prison. He was forced to do hard labor and he experienced great hardship. When I spoke with him after he was released from prison, I asked him what he had found most difficult during those 18 years, and what dangers had he faced. He mentioned that the biggest danger he encountered was the risk of losing compassion toward the Chinese. So he considered losing compassion, or developing anger toward the Chinese, to be dangerous. That is remarkable! This monk is an example of someone who—as part of his mental practice—deliberately keeps a compassionate attitude toward the very people who subject them to great pain. As a result, the mental state of the monk was quite stable. This situation has been researched by scientists, who looked at the mental state of former prisoners in different countries. They found that usually people who pass through such traumatic experiences have an unstable mental state. But in the case of some Tibetan monks, they found that their mental state remained remarkably peaceful, quite normal, in spite of painful circumstances. So everything depends on your mental attitude and not on external events. Crisis or no crisis, money or no money, you can be happy.

Roger Dassen: We have indeed seen a phase in which many companies and banks were engaged in the relentless pursuit of shareholder value, maximizing the P of profit, while ignoring the Ps of people and planet, let alone the P of *pneuma*, as Ruud Lubbers mentioned. These same banks paid lip service to corporate social responsibility (CSR) and sustainability in nice reports, but these were mostly seen as a nice-to-have, and an add-on, without making them an integral part of business and governance. The problem is not so much our ability to monitor the profit side of things, but that we have not found ways to monitor and account for the other Ps of people and planet. We now need to develop instruments to measure these Ps in the same rigorous manner as the P of profit. I believe there will soon become a time that it is no longer an exception that the accountant will raise questions about the other dimensions of business, simply because that is what stakeholders will be demanding from business. There is the encouraging example of the South African stock exchange that makes sustainability a mandatory measure for companies listed on the exchange.[4] This has empowered the accounting industry in South Africa to check companies for their sustainability records. Through examples like this, CSR and sustainability will become the norm rather than the exception.

6.4 Global well-being: learning from the East

Princess Irene van Lippe-Biesterfeld: To my mind, the economic crisis is a reflection of a flawed assumption in Western culture that we need economic growth at all cost, and that greed is good. Your Holiness, may I ask you, you are traveling the world, you are respected in both the East and West, what would you like to tell the Western civilizations at this time of human history, and especially on the subject of compassion versus greed? It seems that in the Eastern philosophical traditions there is a stronger emphasis on compassion, holistic thinking, and balance—which all lead to enhanced well-being—than in the modern-day Western world. Is there something we can learn from the East in this time of crisis?

The Dalai Lama: First, I must say that I see no real distinction between East and West—of course at the secondary level there are differences in culture, habits, and shapes, but at the fundamental level we are just human beings.

4   In 2010 the Johannesburg Stock Exchange decided to require listed companies to adopt integrated reporting, which is referred to in this manner: "A key challenge for leadership is to make sustainability issues mainstream. Strategy, risk, performance and sustainability have become inseparable; hence the phrase 'integrated reporting' which is used throughout this Report." Institute of Directors of Southern Africa (2009). King report on governance for South Africa 2009. Johannesburg: Institute of Directors of Southern Africa. Retrieved from https://jutalaw.co.za/uploads/King_III_Report/#p=1.

PHOTO: LAMBERT VAN DER AALSVOORT

From left to right: the author, Ruud Lubbers, the Dalai Lama, Princess Irene van Lippe-Biesterfeld, and Herman Wijffels.

Mentally, physically, emotionally, we are completely the same. Second, what is clear to me is that all human beings have a need for both material and spiritual progress. Despite economic progress, poverty still exists at national and global levels, which represents not only a moral but also a societal and practical problem. Simultaneously, as the world population increases, there are resource and ecological challenges such as water scarcity and, ultimately, global warming. These are not individual nations' problems, but the problems of 6 billion human beings. So these problems demand a sense of global responsibility and a concern for the well-being of others.

These insights are supported by the scientific recognition that we are social animals. Our individual prosperity and happiness depend on other people. For instance, if the European Union wants its economy to prosper, it can only do so with the rest of the world by, for example, increasing trade. The prosperity of the rest of the world is the basis of your own prosperity. This is the modern economic reality. Whether you call this socialism or capitalism, or East or West, is irrelevant. For this reason, I always explain the value of compassion—not as a religious concept, but simply as a prerequisite for the well-being of us as individuals and the well-being of 6 billion human beings.

6.5 Leadership from science: the promise of neuro-economics

Anders Ferguson: For the last 50 years, you've worked with leading scientists in exploring the intersection between Buddhist science, neuroscience, behavioral psychology, health, and medical science. What has come out of this work is very impressive. We are now getting a much better understanding of the relationship between health, well-being, and the mind. This is important, as it has been revealed that one of the largest diseases afflicting Westerners today is mental depression. At the same time, we're experiencing a global economic depression. So I was wondering, is there a link between the inner and the outer—our inner state of mind and the outer economy, inner and outer depression, so to speak? Here is my question: given so much of your work has been through inquiry with scientists about our inner state of being, and given the global economic and sustainability crises we are now facing, can you imagine now working with economic experts and exploring a new field of neuro-economics? This would be a dialogue focused on the neuroscience of our economy. You could help us inspire such a dialogue.

The Dalai Lama: Sometimes I describe myself as a scientist. In my tradition of Buddhism, we're not reliant on words. We are not even reliant on Buddha's own word. Rather, we rely on critical investigation. That is the true scientific approach. When we explore reality, we should try to remain skeptical, not to accept a belief easily, and continue to investigate and experiment! Then, once something is confirmed, then you can accept it. Buddha himself stated, "All my followers should not accept my teaching out of faith or out of devotion, but rather through their own investigation and experiment."

My interest in science started in my childhood, out of my own curiosity, and some 30 years ago I started to actually meet with scientists. At the beginning, some of my American friends, supposedly Buddhist friends, give me a warning: science is the killer of religion—you'd better be careful. But I thought that in the Buddhist tradition, particularly the tradition that we inherited from the ancient Nalanda University in India, it is customary to study through investigation. If a fact is confirmed through your own investigation, your own critical reasoning, only then should you accept it. Therefore, I thought, as a Buddhist there should be no difficulties in discussing my views with scientists. When I had such meetings, it gradually became very clear to me that the dialogue had mutual benefits. I learned many useful things from modern science. At the same time, some scientists, particularly in the field of psychology and neuroscience, also learned new things. Through our exchange, these scientists obtained a new perspective on their own field. Now these days, there's much more interaction between modern scientists and Buddhist scholars and practitioners.

Now to your question. My answer is: I don't know! I think it would be good to create some occasions where we can put these questions to those who are really trained in the science of neuro-economics and related fields. I'm not such a scientist. I usually consider the definition of science as a technique to investigate reality. So with regard to the economic crisis, obviously further investigation is much needed, objectively, in an unbiased way. In order for it to be a genuine investigation, you must be free from expectations or desires with regard to outcomes. You need to have a neutral and open mind. For example, I'm Buddhist, so when I carry out some investigation, I should not even consider myself as a Buddhist, but as a neutral observer with an open mind! Anything can be possible. Even if we find something that is contradictory to Buddhist teachings, then I should accept that possibility! I think that is the genuinely scientific approach.

I agree with you that it would be very helpful to scientifically investigate what went wrong in the economy, what are the causes of the crisis and how we can avoid one in the future, and to relate all that to how we think, our mental models and attitudes. The same applies to investigating how we can achieve sustainability and how business and economics can help with that too.

I would like to clarify: some people use the word dialogue between Buddhism and science. I think that is not correct.[5] Broadly speaking, Buddhism is divided into three parts. First, there is Buddhist science, which is simply trying to know reality and includes our mind, our emotion, our body, and all external things. Second, on that basis, there is Buddhist philosophy. When you have some knowledge about reality on a deeper level, such as what we in Buddhism call "emptiness" or "interdependence," you see that with our ordinary mind we don't perceive that reality. Ordinarily, we misperceive reality because we have some sort of distortion in our mind, which we call ignorance, which is in fact the reason why we are often unintentionally creating suffering for ourselves and others. This understanding leads us to explore the possibility of changing our mind, our perception, so

5   For a deeper discussion of the Dalai Lama's view on this point, see Dalai Lama (2005). *The Universe in a Single Atom*. New York, NY: Random House.

that we remove our misperception and learn to perceive reality as it is. This brings us to the Buddhist concept of elimination of all misconception, taking away the causes of suffering, which leads us to experience happiness. So that's the second part, Buddhist philosophy. Third, on the basis of the first two parts, there is Buddhist religion, including meditation. If we practice meditation, we can improve and develop our mind so that we eliminate all causes of suffering and create causes for happiness.

What I am saying is that our meeting with science only concerns the part of Buddhist science, not Buddhist philosophy or religion. For example, we never talk with scientists about subjects such as reincarnation or karma. I'm always telling them—this is the business of Buddhists, not the business of scientists. So our common interest with science is simply the question: what is reality? In many fields, the explanations provided by modern science are much more advanced than that of Buddhist texts, such as in the field of quantum physics. In Buddhism we had concepts about very small particles that make up matter, but Western science has, through experimentation, provided confirmation and greatly deepened our understanding of these subtle levels of reality. However, with regard to the mind, Western or modern science is still quite young compared to ancient Indian psychology. There is a lot of information from ancient Indian sources about the mind and emotion, which can be very useful for the Western scientist.

Sometimes I feel that Buddhism can serve as a bridge with science on behalf of all religions. Theistic religions such as Christianity and Islam do not seem to be interested in exchanging ideas with modern scientists, while many modern scientists have no interest in religious spirituality. Buddhism can perhaps be a bridge between these two. As I said, Buddhist science agrees with modern science in simply trying to understand the nature of reality. At the same time, Buddhist concepts such as *nirwana* that arise from deeply understanding reality can be helpful for understanding other religions' concepts such as salvation or heaven.

Sander Tideman: I am aware that in Zurich next year, His Holiness will participate in a Mind and Life Institute conference on economics with a focus on compassion in economics, so we can deepen this inquiry at that meeting.[6] Perhaps for the audience here, the majority of whom are working in business and society and not so much in science or philosophy, it

---

6   Singer, T., & Ricard, M. (2015). *Caring Economics: Conversations on Altruism and Compassion Between Scientists, Economists, and the Dalai Lama.* New York, NY: Picador.

would be helpful to hear your views on what it is that you regard as the most important outcomes of the exchange between Buddhism, psychology, and neuroscience? What have you learned from these dialogues?

The Dalai Lama: One of the most important outcomes is a common understanding that love and compassion have profound effects on health. Emotional and physical stability in adult life are determined to a large extent by the care and affection each of us receives from our mothers; love and affection sustain many physical functions and particularly the immune system, as well as having an obvious importance in human relations. Hatred and anger have deeply destructive effects on one's own body, as well as alienating and harming others. Based on these considerations, it is clear that ethics can be seen in a biological context, and as such need not be associated with specific religious concepts. Indeed, as such a huge proportion of the world's population has little or no deep religious interest, it is important to establish a system of "secular ethics" to complement ethics based on religious belief, whether theistic or (like Buddhist practice) nontheistic. A recognition of the biological benefits of having a compassionate, warm-hearted approach to others, and the recognition of others as having feelings and rights no less important than one's own, can form a powerful basis for the development of such secular ethics.

Furthermore, this approach can and should be applied by all of us at an individual level. It is here that I believe business leaders have a responsibility and an opportunity to create positive change in the world. It is no use waiting for the building of a more compassionate society in any other way than each of us applying these ideas first in our own minds, then in a family context and in a business context, and then in wider and wider circles like the ripples on a pond. No one, I believe, should feel he or she cannot contribute.

6.6 What can you do as an individual?

Li Hong Hui: Your Holiness, I'm from China, but many years ago I moved to Holland where I work and live. When I was in China I heard so many terrible stories about you, from my primary school, from my parents, from my high school, from my university. But when you came in, when I saw you, I saw nothing terrible: I saw compassion; in fact I felt pervasive compassion.

And suddenly all those stories dissolved. So thank you for giving me such a wonderful experience.

The Dalai Lama: Thank you!

Li Hong Hui: My question concerns the paradox of spiritual practice. On the one hand, we have the tradition of cultivating our own personality, our own compassion, our own inner life, which can be such a blissful experience when we are doing our own meditation at our home, solitarily, or with a few friends. On the other hand, I know the purpose of practice is to bring that spirit or consciousness into life, so that we can touch life. My question is: how can we bring our heart fully into life, rather than having only the spiritual bliss of having peace of mind?

The Dalai Lama: First, you're quite fortunate, meeting with a demon. Because this is how the Chinese media describe me: a demon. Recently, an American friend asked me: why is the Chinese government afraid of you? And I answered jokingly: because of my horns, the horns of a demon.

I really appreciate your explanation. We Tibetans and Chinese brothers and sisters have had close relations for thousands of years. Sometimes fighting and quarreling, and at one point in the seventh century the Tibetan army invaded China. Recently, it turned the other way around. We must resolve the problems between the Tibetans and our Han brothers and sisters.

When I give lectures in America and Europe, young Chinese students who attend my lectures often approach me afterwards. They tell me that when they were in mainland China, they thought the Dalai Lama was very much an orthodox religious person, with a closed mind. When they listen to my talks, they are often very surprised and change their opinions about me. Many try to meet me in person. As they are still very much afraid of the Chinese government, these meetings need to be in secret. Here too you can see, as we discussed, that it is most important to see reality as it is, regardless of whether it is good or bad, through our own explorations, not because of what people say or believe.

Now to your question. What is the real purpose of our meditation? First, it is not only for the period of the meditation, like having a blissful experience, but something for the rest of our day. It is like an electrical recharge. The very purpose of

> Our meditation is meant to shape our mind, to shape our attitude—the attitude toward problems, toward business and the economy, and toward ecology.

the recharge is the use of it throughout the rest of the day. Our meditation is meant to shape our mind, to shape our attitude—the attitude toward problems, toward business and the economy, and toward ecology.

For example, when you meditate on compassion for, let's say, half an hour, then some impact of that meditation remains in your mind. So during the rest of the day, when you are meeting people, the impact from the meditation can help to sustain your peace of mind. This is particularly the case when you're facing difficulties—a compassionate mind is really helpful in sustaining your peace of mind.

This is much more effective than developing anger or aggression in the face of problems. There's a Tibetan saying: if the mind is a peaceful mind, you can use your intelligence properly. When your mind is relaxed, you can be more intelligent and this will help you to be more effective in your business and in your life.

Developing the right attitude is the purpose of meditation, or the purpose of religion for that matter. I always stress that whether we accept a religion or not, that's up to the individual. But once we accept religion, then we should be serious and sincere. The religious teaching should be part of our daily life. So that when you are carrying out your life, the impact of your belief is expressed in your day-to-day actions. When you are doing meditation, you may look very calm, but then as soon as you experience a problem, you forget about the meditation—I think that is a mistake.

The ultimate purpose of spiritual practice is to build a compassionate world. How? A compassionate world may sound too ambitious. One individual can only have a very small impact. In reality, any change that happens in the world, good or bad, actually comes from individuals. First as an individual, try to develop inner peace on the basis of compassion. Then after that, try to create a more compassionate family, then another ten compassionate families, and another hundred compassionate families. This is the proper way to develop a better society, a better humanity. It takes time, we need constant effort, courage, and determination. Some obstacles, some difficulties are bound to encountered. Regardless of these obstacles, as your goal and your motivations are right, don't think that achievements must be realized within one's own lifetime. Don't think that way.

We must start here and now to build a compassionate society. We may not see the results, but our effort is not wasted. Our generation can start with awareness, and then the next generation will follow with the next steps. Then, perhaps at the end of this century or the next century, we may achieve a more compassionate world.

When we became refugees we faced big challenges, and so we made a slogan: "Hope for the best, but prepare for the worst." This slogan helped us to pass through the difficult years in exile with determination and some success. I don't think that when the Chinese forces crushed Tibetan resistance in 1959 they expected the Tibetan spirit to be still alive 50 years later; the same spirit we should now carry forward in building a compassionate society. For Tibetans, that means cooperating with our Chinese brothers and sisters, not splitting our country from China, but building a modern economy with genuine autonomy within the larger Chinese State. I think that is possible.

Similarly, I think it is possible to build a compassionate society and sustainable economy through small but determined steps. It seems a large and complex undertaking, but each one of us can do something, in our family and our company. Step by step. Thank you.

Part 4:
Education of the heart

7
Living shared purpose

This chapter describes my work after the 2009 dialogue. As the financial crisis worsened, it first turned into the Euro-crisis and later into a fully fledged economic crisis. The whole ideology of free-market capitalism was called into question by many observers. While most governments sought to combat the symptoms with changes in financial structures, I felt the need to look at the human and social dimension—the human "how" of change.

Based on the insights derived from the public dialogues with the Dalai Lama, I started to work on a framework that would help leaders to create genuine sustainable value, at personal, organizational, and societal levels—or "triple value." In order to facilitate my research, I partnered with academic and educational institutions and experimented with various educational designs. Exposure to high-performance teams in business and sports allowed me to conceive of new educational approaches to achieving sustainable value creation and solving global challenges.

Working with young and upcoming leaders in various fields on topics that mattered to them and the world gave me a deeper sense of fulfillment. This phase presented the fourth stage in my quest for making a difference: "living shared purpose." By now I could literally share my purpose with others. This phase culminated in the "Education of the Heart" symposium, the fourth "Compassion or Competition" dialogue which the Dalai Lama once more supported.

7.1 The "how" of change

The "Leadership for a Sustainable World" conference with the Dalai Lama in 2009 coincided with the unfolding and deepening economic crisis—from the credit crisis to the Euro-crisis and to the collapse of entire national economies. The newspapers were full of serious questions about the neoliberal style of capitalism that had dominated the last few decades. What transpired in the postcrisis world was a growing understanding that, without a new form of capitalism, modern society will continue to suffer from unpredictable and multiple (economic, societal, and ecological) crises that erode principles of sustainability, equity, and societal harmony. The big question is no longer whether the system needed to change, but *how* it can be changed. Most economic and political authorities started to amend the system through structural adjustments: bailing out banks, higher reserve ratios, strict monetary controls, cutting costs, monetary easing, stimulating consumer spending, and so on.

But I was among many who felt that there was something missing in this focus on financial structures. While our political leaders spoke of monetary reform, debt restructuring, and austerity, I started to see a much deeper crisis in the Western world: a crisis of values, worldviews, and vision. On what basis are we making the choices for restructuring our financial sector? While banks were saved because they were "too big to fail," many bank executives escaped punishment for bringing their banks to the brink of collapse, often landing with a golden parachute, which is in stark contrast with the many who lost their jobs as a result of the economic crisis. Since this challenged people's sense of fairness, it eroded public trust in both finance and government and added to the dwindling trust in the economy. This resulted in a downward, vicious cycle.

To make matters worse, at least in Europe, government leaders failed to communicate a vision of how austerity can contribute to innovation, entrepreneurship, and creativity for building a sustainable economy. While taxpayers' money was used to bail out failing financial institutions, unemployment figures shot up and educational budgets were cut. This was compensated by quantitative easing,[1] but this caused little relief. In fact, it caused long-term interest rates to reach record lows, in some cases even

---

1   Quantitative easing is the purchase of government bonds using money held by central banks—or, in plain language, the printing of money by the central banks.

turning negative, without substantially invigorating consumer demand. To make it worse, it increased public debt in the U.S.A. from 64% of GDP in 2008 to 104% in 2014, while in Europe it rose from 66% to 93% and in Japan from 176% to 237%.[2] In other words, we were treating the economic crisis by stimulating debt. This was like treating a sickness by taking the poison that caused it!

It seemed that monetary restructuring had become a goal in itself, without any overall view of what creates a sustainable economy or resilient society. What was missing in the focus on structural change, to my mind, was an understanding of what underpins sustainable societies and economic systems, namely human beings. Specifically, we were ignoring the human capacity for resilience, initiative, optimism, responsibility, and creativity—in short: the human "how" of change.

In spite of the doom and gloom that spread at macrosocietal levels, I was encouraged by what I learned from my own work in leadership development: namely, that many business organizations were embracing sustainability and the power of social networks as sources of business innovation. Thanks to the crisis, many business leaders started to see that "business as usual" was no longer an option. The sudden rise of companies such as Uber and Airbnb stimulated business model innovation by using opportunities provided by new technology in particular. There was also an uptick in consumer awareness of the need for a more sustainable, circular economy, which encouraged companies to develop new products and services. Investors began to look for more in-depth information about companies' environmental impacts and there was a growing recognition by some regulators that a voluntary approach to reporting will not deliver the changes required to move to a more sustainable economy.

As the credit crisis put limits on the traditional method of leveraging financial capital as a means to create profit, many businesses started to focus on leveraging **human and social capital**. On the downside, this gave rise to the unhealthy and unfortunately widespread picture of overworked employees, many of whom had to lead a "divided life," leaving their values at home when they went to work. However, the upside was that companies discovered that if employees' values are left at the door of their professional life, then the enterprise ultimately loses—and so does society. Rather than treating people as "human capital" or "human resources," companies

2   The Economist (2016, February 20). The world economy: out of ammo? *The Economist*, 36.

started to treat people as a *someone*, not a something, as a *whole person* with untapped human potential, not a mere employee offering their time.

Respecting the whole person includes thinking of people in all their various roles in relation to the business: as employees, customers, suppliers, investors, and citizens. In this view it is understood that people are driven by a sense of purpose, seeking outcomes that enable people to reach their full potential. It means contributing fully to building relationships within the workplace and the outside world, which can engender trust between people and between business and society. Even though the reorientation toward developing human potential is still difficult in many business settings, it is no longer a viable excuse to say that people cannot and do not want to change. In other words, business is discovering the human "how" of change.

7.2 Looking at the new reality with outdated concepts

These developments strengthened my resolution to develop a coherent intellectual framework for leaders to respond adequately to the crises, by creating economic, societal, and ecological value. The crisis gave me the courage to make this the prime focus of my work. I recognized the societal needs for it and by now I had gathered sufficient experience and knowledge. I now shifted into a phase of actually living shared purpose.

The Dalai Lama already said at the first dialogue in 1999 that, in order to deal with complex global challenges, we need to expand our narrow short-term perspective into a wider long-term view. Ten years later, in the 2009 dialogue, he made the following comments: "the current financial crisis could teach us to look beyond material values and unrealistic expectations of limitless growth. When things go seriously wrong, it is often because a new reality is still being viewed with outdated concepts, and this is certainly the case with the economy today."

I took these words as an overarching perspective for my work. Since the dialogues had revealed that changes occur at roughly three levels, from the macro, meso, to micro level, I felt that the new conceptual model also needed to contain three levels:

1. The larger system (including globalization, macroeconomics and economic policy, society, ecosystems)—which I called the "All" dimension

2. The business organization, and its products and services that are offered in the market—which I called the "Org" dimension

3. Leadership, individual mind-sets, views, intention, responsibility, ethics, and efforts—which I called the "I" dimension

By studying the changes at these levels, I concluded that at each level we were witnessing a similar shift in paradigm. This is a shift from the **growth–transactional** paradigm to the **sustainable–relational** paradigm, which entails an expansion of value creation from purely singular financial value for the benefit of the shareholders, to multiple value creation, including societal, ecological, and economic value for the benefit of all stakeholders— "Triple Value" (this is further explained in Chapter 9). Paradigms are the beliefs that we hold, unconsciously shaping the way we perceive and relate to the world. I found that, through reflective practices such as mindfulness and dialogue, people could become aware of these mind-sets and gradually shift from outdated restrictive mind-sets to more current constructive mind-sets. This exploration culminated in a book called *Mind Over Matter*, in which I argue for a new paradigm for leadership in business and economics.[3] While this was still an evolving piece of research, it caused the management of Nyenrode Business University to offer me a research position for continuing the project. The specific objective was to make the emerging insights relevant to mainstream business leadership.

7.3 Compassionate competition

Then, in 2010, as the first exploratory phase of the research, I set out with my research partner Muriel Arts to conduct in-depth interviews with leaders in business and sustainability, and we held group dialogues with executives of companies such as Unilever, DSM, ING Bank, Nike, Deloitte, and Hay Group. We found considerable convergence among both business thinkers and practitioners on both the analysis of what had gone wrong in capitalism and on the way forward, transforming capitalism. The many ideas that we gathered resonated with the guidelines that we had received in the dialogues with the Dalai Lama. Harvard University's business scholars Michael Porter and Mark Kramer describe the situation as follows:

3   Tideman, S. (2009). *Mind Over Matter: Van Zeepbelkapitalisme naar Economie met een Hart*. Amsterdam: Business Contact Press.

> In recent years business increasingly has been viewed as a major cause of social, environmental, and economic problems. Companies are widely perceived to be prospering at the expense of the broader community ... A big part of the problem lies with companies themselves, which remain trapped in an outdated approach to value creation which has emerged over the past few decades ... Companies must take the lead in bringing business and society back together. The recognition is there among sophisticated business and thought leaders, and promising elements of a new model are emerging.
>
> Yet we still lack an overall framework for guiding these efforts, and most companies remain stuck in a "social responsibility" mind-set in which societal issues are at the periphery, not the core ... The solution lies in creating economic value in a way that also creates value for society by addressing its needs and challenges; business must reconnect company success with social progress.[4]

The objective of our research, then, became to integrate these converging insights into a framework for business leadership to create shared value. In this way, companies would serve both the needs of society and themselves as an expression of "enlightened self-interest," thus putting the apparent tension between compassion and competition to rest.

Perhaps this would lead us to conclude—as the Dalai Lama had suggested all along—that the best way to compete is by developing compassion. This would be a radical departure from conventional business thinking and there would be fundamental implications for business education. Specifically, if we want to truly create a more sustainable economy and flourishing society in addition to certain structural changes, we will need to cultivate different human mind-sets, values, and behaviors. This applies particularly to our leaders and managers who seem to approach the crises with outdated conceptual models and flawed values, such as greed and lack of transparency.

In 2010, at the Mind & Life Institute conference in Zurich, I heard the Dalai Lama emphasize the need for leadership development: "Leaders are not born or made, but cultivated and developed. The best way to develop leaders is through the training of values."[5] The development of value-based leadership, then, is the human "how" of change, beyond the mere structural approach to which most leaders currently subscribe.

4   Porter, M., & Kramer, M. (2011). The big idea: creating shared value. *Harvard Business Review*, 1 (January–February), 64-77.

5   Singer, T., & Ricard, M. (2015). *Caring Economics: Conversations on Altruism and Compassion Between Scientists, Economists, and the Dalai Lama*. New York, NY: Picador.

Part of my research was working directly with leaders. By serving on the faculty of Nyenrode's Executive Education and Windesheim Honors College, I could test evolving ideas and practices on executives and students in classroom settings. These included practices such as mindfulness and reflective dialogue, for which there was an increasing appetite in management.[6]

> Leaders are not born or made, but cultivated and developed. The best way to develop leaders is through the training of values.

I also continued to run custom-made executive programs for companies and lead yearly leadership journeys to Asia. After China restricted foreign travel into Tibet, I chose other destinations instead. Most special was Bhutan, thanks to its secluded location, pristine nature, and authentic Buddhist culture. It proved to be the ideal setting for leadership transformation at the three levels of "I," "Org," and "All." The country's development concept of gross national happiness served as the all-level backdrop for leaders to explore their own assumptions about happiness and the purpose of their work and role in society. It was heartwarming to see people, while enjoying beautiful landscapes, discover new sources of inspiration. When people learn to let go of restrictive mind-sets and adopt more expansive and holistic frames of reference, they often discover new inner resources that empower them to make life-changing decisions. For some, the change was confined to their personal life, while others were able to impact the organizations that they work for. It was a delight to witness these manifestations of "shared purpose."

7.4 High-performance sustainable teams

I could observe the same principles at play much closer at home. The most surprising place was the world of soccer. Raymond Verheyen, a seasoned

---

6   Mindfulness has been adapted from Buddhist meditation and applied in a secular medical context by Dr. Jon Kabat-Zinn. His program at MIT, called "Mindfulness-Based Stress Reduction," proved to be so successful that they set out to train hundreds of people as mindfulness instructors, who gradually introduce it into business settings. See, for example, Kabat-Zinn, J. (1990). *Full Catastrophe Living: Using the Wisdom of Your Body and Mind to Face Stress, Pain, and Illness.* New York, NY: Bantam Books. For the application of mindfulness in business, see: Hougaard, R., Carter, J., & Coutts, G. (2015). *One Second Ahead: Enhance Your Performance at Work with Mindfulness.* London: Palgrave Macmillan.

physiologist who had worked with various top teams in the world and was the founder of the World Football Academy, asked me to train soccer coaches in mindfulness and leadership. As part of my preparation, I delved into the latest science of health and top performance in sports, where—to my surprise—I found insights that I could apply directly to sustainable business leadership. After all, the underlying paradigm was the same: human beings are complex adaptive "living systems" that operate on the basis of mutually interdependent relationships, with the mind being an integral instrument for realizing well-being and sustainable performance. Through Raymond and the participants in my program at the World Football Academy, I was exposed to what the world's best teams are doing to ensure that their players and teams remain ultra-fit and resilient. The case of FC Barcelona, which was by far the most successful European soccer team of the last decade,[7] was most impressive. I discovered the secret that turned the club into the second most valuable sports team in the world, worth US$3.16 billion, and the world's second-richest soccer club in terms of revenue, with an annual turnover of €560.8 million.[8]

What is the secret? Even though the most immediate goal of soccer, quite literally, is to score goals, the philosophy of Barcelona is much more far-sighted and profound. "Scoring a goal is just one intermediate result in an elaborate chain of causes and effect, starting with the values, beliefs and dreams of FC Barcelona players, staff and fans," said Albert Capellas, a coach at the FC Barcelona Youth Academy. He explained that all players are trained to express the values of the Catalan people in their attitude and playing style, based on a culture of solidarity, creativity, and fun. FC Barcelona has summarized these values into three core principles of "fun, run, and play," on which the club established its entire scouting, training, and playing methodology. Fun stands for a positive optimistic mind-set capable of overcoming setbacks, run for commitment and hard work, and play for

[7]   In international club soccer, Barcelona has won five UEFA Champions League titles, a record four UEFA Cup Winners' Cups, a shared record five UEFA Super Cups, a record three Inter-Cities Fairs Cups and a record three FIFA Club World Cup trophies. Barcelona was ranked first in the IFFHS Club World Ranking for 1997, 2009, 2011, 2012, and 2015. For FC Barcelona's impressive record, see https://www.fcbarcelona.com/club/history/card/fc-barcelona-team-records.

[8]   Deloitte Sports Business Group (2016). *Top of the Table: Football Money League*. Retrieved from http://www2.deloitte.com/content/dam/Deloitte/uk/Documents/sports-business-group/uk-deloitte-sport-football-money-league-2016.pdf.

collaboration and compassion. Players exhibit these values not only to win the game, but also to engender these values among the club's supporters, including society at large. It is for this reason that FC Barcelona for many years refused to sign a commercial shirt sponsor, instead displaying the UNICEF logo free of charge. In addition, the FC Barcelona Foundation provides €2 million annually to the UNICEF partnership in order to implement projects to combat HIV/Aids and to foster children's education through sport in the most needy areas of the world.

This deal with UNICEF is not just an act of charity, said Capellas, but part of Barcelona's enlightened self-interest. By serving a societal goal at the "All" level, the spirit of generosity of the players is also unleashed at the team level ("Org"). There is no Barcelona player who is not molded by the frame of "fun, run, and play," as well as serving the higher purposes of "Org" and "All." Even a top player such as Lionel Messi is aware of his interdependence with others, which he expresses in his sportsmanship. When he celebrates a goal, he first thanks his fellow players and the fans. To put it differently, even though individual players excel at the "I" level, they are visibly motivated by a shared purpose with others, which resonates throughout the entire organization. Shared purpose is also reflected in the structure of the organization: unlike many other soccer clubs, the supporters own and operate FC Barcelona.

The secret of the FC Barcelona model is that its philosophy has been made through an explicit shared purpose, which is expressed in clear rules and behavior. While this purpose is rooted in traditional Catalan culture, they came to life in soccer when the Dutch star player Johan Cruyff brought the concept of "Total Football" to Barcelona in the 1970s,[9] which at that time suffered from repression under the Franco regime. The Total Football concept, fused with Catalan values, gradually turned into a philosophy for sustainable high performance.

FC Barcelona's model could be seen as an expression of "shared purpose" aimed at creating "triple value." I realized that soccer could help me on my quest for developing higher-purpose leaders for sustainable higher performance.

9   Martínez, R. (2010, July 11). World Cup final: Johan Cruyff sowed seeds for revolution in Spain's fortunes. *The Daily Telegraph.*

7.5 High-performance sustainable organizations

I started to use the example of FC Barcelona and other sports to show businesses what it means to build high-performance teams. Using the same underlying principles, leaders could then discover for themselves the pathway to building "high-performance sustainable organizations" (HPSOs). By defining an HPSO as an end-point of organizational development, where the entire organization is driven by a shared purpose for creating "triple value," we raised the bar for business leadership. The example of a top performing team such as F.C. Barcelona helped them to understand that sustainability is not just a complex technical fix for abstract challenges, but an expression of a particular culture and style of leadership. Both sustainability and top performance are manifestations of shared purpose in action—generating fun, joy, and success.

A business leader who embodied such shared purpose-based philosophy is Paul Polman, who became the CEO of Unilever in 2009. Shortly after joining Unilever, he launched the Sustainable Living Plan in 2010, with the explicit purpose "to make sustainable living commonplace."[10] The plan aims to double the size of the business while reducing Unilever's environmental footprint and increasing its positive social impact. Unilever wanted to change things on a global scale: deforestation and climate change; water, sanitation, and hygiene; and sustainable agriculture and smallholder farmers.

This ambitious plan represented an "All" purpose behind the business. It was no longer enough for Unilever to create excellent financial value for the company, but instead its aim became to solve critical societal challenges. While a shared purpose was already implicit in Unilever's culture, as I had learned from my work with Unilever's brands Lifebuoy and Ben & Jerry's (Chapter 5), Polman made the company's shared purpose explicit and connected it to ambitious yet tangible goals, just as Barcelona's coaches did in soccer. This is evidenced by the following words from Polman:

> I know we all have our jobs, but that has to come from a deeper sense of purpose. You have to be driven by something. Leadership is not just about giving energy but it's unleashing other people's energy, which comes from buying into that sense of purpose.

10   Unilever (2016). Sustainable living. Retrieved from https://www.unilever.com/sustainable-living.

But if that purpose isn't strong enough in a company, if the [leadership] doesn't walk the talk, then the rest will not last long. The key thing for CEOs is to make that a part of your operating model.[11]

And Polman walked the talk. Via the Sustainable Living Plan, he decoupled economic growth from carbon growth across the entire business. Most boldly, he refused to report quarterly to Wall Street analysts, thus challenging the orthodoxy of managing on a 90-day payback period.

Similar to FC Barcelona, this attitude did not distract the company from achieving business success, but instead invigorated it. Each year, from 2011, it recorded volume growth ahead of the markets, steady improvement in core operating margin and strong cash flow. Over this period, its growth in dividends per share increased 6.14%.[12]

Companies such as Unilever are smartly leveraging a trend of purpose-driven behavior in society in general—a verifiable wave of health-conscious consumers, meaning- and balance-seeking employees, future-focused students, and increasingly influential faith-based and spiritual communities.[13] Purpose, of course, is a natural ally of sustainability—at a personal level, at an organizational level, as well as at a societal level. There are a number of global trends supporting this, including at an international political level.

> Purpose, of course, is a natural ally of sustainability—at a personal level, at an organizational level, as well as at a societal level.

The new UN Sustainable Development Goals (SDGs), first introduced as a proposal at Rio+20 in June 2012, have been formally launched to replace the Millennium Development Goals.[14] The SDGs can serve as an empowering basis for innovation at many levels, and there is already an impressive pool of expert leaders that have embarked on SDG-driven journeys. Companies such as Unilever, Novozymes, Whirlpool, and BASF have been vocal about aligning brand strategies, product and service innovation pipelines, supply chain

11   Confino, J. (2013, October 2). Interview: Unilever's Paul Polman on diversity, purpose and profits. *The Guardian*. Retrieved from http://www.theguardian.com/sustainable-business/unilver-ceo-paul-polman-purpose-profits.
12   Financial Times (2016). Equities. Retrieved from http://markets.ft.com/research/Markets/Tearsheets/Financials.
13   Williams, F. (2015). *Green Giants: How Smart Companies Turn Sustainability into Billion-Dollar Businesses*. New York, NY: AMACOM.
14   United Nations (no date). Sustainable Development Knowledge Platform. Retrieved from https://sustainabledevelopment.un.org.

initiatives, employee engagement programs, and overall corporate sustainability goals with the SDGs.

In her book *Green Giants: How Smart Companies Turn Sustainability into Billion-Dollar Businesses*, Freya Williams explores what underlies the success of the world's first nine billion-dollar sustainability-focused brands—Nike, Unilever, Chipotle, Toyota, Tesla, GE, IKEA, Natura, and Whole Foods.[15] These companies are shattering the myth that acting sustainably and building a billion-dollar business are mutually exclusive. Williams identifies six classes of success factors:

1. Iconoclastic leadership fueled by deep conviction and a rebellious streak

2. Disruptive innovation that uses sustainability to spur the development of radically better products and services

3. Mainstream appeal with positioning and packaging stripped of the crunchy clichés that alienate the average customer

4. Embedding sustainability values throughout the entire organization

5. Establishing a new "behavioral contract" with consumers and other stakeholders

6. Most importantly, a higher purpose that ignites the company

7.6 The role of management education

Excited by these examples of purpose-driven businesses, I was struck by how little of this knowledge trickled down into mainstream education, especially management education. It was obvious that when people can link their work to a higher purpose, they are more motivated to bring the best of themselves into their work. Why don't we teach the power of shared purpose at business schools? It was at that time that I met Rob van Tulder, Professor of Business and Society at the Rotterdam School of Management, Erasmus University,

---

15 Williams, F. (2015). *Green Giants: How Smart Companies Turn Sustainability into Billion-Dollar Businesses*. New York, NY: AMACOM.

which is the leading business school in the Netherlands.[16] Rob was intrigued by our research and decided to collaborate on it. Soon after, we obtained a major grant for our research project from NWO, the Netherlands Scientific Research Agency, and we moved the entire project to Erasmus University. It turned out that we also shared an interest in integrating sustainability into management education. When he learned about my relationship with the Dalai Lama, we decided to invite the Dalai Lama to visit Erasmus University to explore these questions, particularly how we can bring concepts such as "shared purpose" further into education.

I felt that the call for purpose-based education really summed up the essence of the Dalai Lama's message for the Western world. Concepts such as purpose, compassion, and responsibility are the core of what he stands for. In his book *Beyond Religion* he writes:

> My hope and wish is that, one day, formal education will pay attention to what I call education of the heart. Just as we take for granted the need to acquire proficiency in the basic academic subjects, I am hopeful that a time will come when we can take it for granted that children will learn the indispensability of inner values such as love, justice and forgiveness. I look forward to a day when children and students will be more aware of their feeling and emotions and feel a greater sense of responsibility both towards themselves and towards the wider world. Wouldn't that be wonderful?[17]

Inspired by this statement, I reached out to educational experts in the world, including the well-known psychiatrist Dan Siegel, who founded the field of interpersonal neurobiology.[18] Dan Siegel, Rob van Tulder, and Muriel Arts helped me to conceive of a next round of dialogues with the Dalai Lama, which we titled "Education of the Heart." The heart represents the human quality to connect with others and the environment, and thus serves as a conduit for (shared) purpose, motivation, responsibility, and care. To begin, we wrote down what we believed to be the underlying problems with mainstream education.

16   van Tulder, R. (2012). *Skill Sheets: An Integrated Approach to Research, Study and Management.* Amsterdam, Netherlands: Pearson.

17   Dalai Lama, & Norman, A. (2011). *Beyond Religion: Ethics for a Whole World.* New York, NY: Houghton Mifflin Harcourt Publishing.

18   Siegel, D. (2009). *Mindsight: The New Science of Personal Transformation.* New York, NY: Random House. See also http://www.drdansiegel.com.

7.7 Education of the heart: toward a new educational paradigm

Current educational practices have largely been designed in and for an era that looks different from today. This was of benefit at the start of the 1900s, but society and its citizens have changed and there is a now a pressing need for an educational system that reflects the age we live in. For example, our changed environment demands a more collaborative approach to education, forcing teachers and stakeholders to develop new curricula together. However, an increasing number of pupils, students, and graduates find themselves rather ill-equipped to face the pressures of modern life, let alone to generate creative solutions to the increasing social and ecological challenges. These challenges are systemically linked to current knowledge generation and education. For over four centuries we have refined and developed our ways of understanding the world based on an ambition for objective and value-free (scientific) knowledge. The school system, which was built to support this understanding, concentrates on developing (disciplinary) brains and rational thinking.

This approach has brought undisputed progress in many areas. It has also caused—as an unfortunate side-effect—us to treat and view our environment as being outside ourselves and a resource to be exploited and consumed. In many respects, this vision has also come to dictate the way we treat each other—as co-workers, as suppliers in supply chains, as rational consumers, or as calculating citizens. As a result, our dominant worldview largely omits to make sense of the "subjective" world, including the connection experienced between humans and our planet that we rely on and are part of.

The prime aim of education has consequently become to develop objective, rational, and value-free knowledge. We "master" and manage our relationships, the world, and ourselves. These cognitive skills may be needed to earn a degree that gets you a job and to acquire a minimum degree of scientific thinking. But these skills have proven insufficient for people to develop all of their capacities. They have not enabled people to understand the complexities of today's society, its opportunities, and its systemic challenges. We have created an artificial separation between mind, heart, body, other human beings, and our natural surroundings. Furthermore, the economic crisis reinforces a tendency to finish degrees in less time, specialize early, and obtain quick results. This creates an artificial separation between schooling, life, and work, even though it is in the integration of these domains that we find the sources of health and happiness.

7.8 Beyond religion: the sustainable-relational mind

Recent insights from neuroscience and psychology, as well as examples from top athletes and high-performance teams, provide a new perspective on the best approach toward education. It reinstates the notion that our subjective and relational experience is an essential element in learning, which in many respects corresponds to insights from ancient wisdom traditions, well described by the Dalai Lama in *Beyond Religion*. What we call rational or cognitive learning (the "head") represents only a fraction of human capacity. The subjective experience, including emotions, social skills, and motivation (the "heart"), represents the much-needed ability to adapt to, and succeed in, the challenges of today.

According to science, our mind turns out to be more than our physical brain: the neural network spreads out all over the body. As Daniel Siegel asserts, the cluster located at the heart is not only an integral part of the neural circuitry including the brain, but serves as a conduit for the activation of feelings such as empathy, a sense of ethics, orientation to social relationships, and generating feelings of well-being and contentment.[19] In other words, the heart is not only relevant as a metaphor for the "soft" or emotional side, but it is also the foundation of emotional resilience, well-being, and happiness, as well as our relationships with others. Moreover, experiments at the intersection between neuroscience, psychology, and mindfulness have shown that this subjective "heart" dimension can be cultivated and trained. In other words, the subjective experience represents a large untapped potential for humanity.

What does this mean to education? What we have relegated to being unscientific (and generally considered part of the domain of religious education) is now being revisited and reinvented. It provides a new perspective on unlocking human potential for tackling the challenges of today. Therefore, this previously "missing" piece in our understanding is an essential part of how we can "upgrade" education for the modern world. This approach to education can no longer be the exclusive domain of religion or ideology: it can and needs to be fully integrated in the discourse of modern science and education. The Dalai Lama describes this as the introduction of "secular ethics" in education.

---

19   Siegel, D. (2013). *The Developing Mind: How Relationships and the Brain Interact to Shape Who We Are.* New York, NY: Random House.

Why is this insight so important at this time and age? When people develop these qualities of the heart, they become more capable of building sustainable relationships, which in turn are the building blocks for sustainable organizations that can contribute proactively to fulfilling necessary societal and ecological needs. The growing societal and ecological needs can only be met when people collaborate and collectively find new solutions. The challenges today are too complex for any single individual, organization, or nation to solve. What is needed, therefore, is a collective creative effort from all sectors in society and the economy. Building a sustainable economy can only be achieved when all people—through their jobs, companies, and network of relationships—bring their full hearts to their work. That is why the education of our heart should underpin business and management education in particular. Business organizations have the means to scale up and to reach many people and communities across the world. In this respect, business has more power than government.

Education of the heart, based on "secular ethics" and our subjective and relational experience, extends beyond individual actions into collective systems. It includes societal health, well-being, sustainability, and ecological resilience. In a sense, we are talking about the education that can ensure the survival of (human) life on this planet. The new approach in education is, in essence, an inquiry into developing the "high-performance sustainable and relational mind." The features of this new education are summarized as three dimensions, which correspond with the three levels of complexity previously identified in the dialogues with the Dalai Lama and the concept of triple value:

1. **Personal ("I"):** the development of all human capacities, not limited to cognitive ability, including the physical–emotional–social–societal dimensions of human beings. In other words, the whole being.

2. **Organizational ("Org"):** emphasis on the need for interactive valuable relationships between employers and employees, between companies and their clients, between students and teachers, and between school, parents, family, community, business, and society. In other words, the relational dimension.

3. **Systemic ("All"):** cultivating an attitude of taking responsibility for the happiness and the well-being of all, i.e., oneself, one's relationships, one's organization, and one's community, society and ecosystems.

On the basis of this analysis, we invited the Dalai Lama to participate in the "Education of the Heart" symposium at the Rotterdam School of Management, Erasmus University. The Garrison Institute in New York provided funding. The symposium brought together a wide range of educational experts and practitioners, including a large number of students and executives. The call for a different approach to education seemed to resonate with many people—within days of announcing the symposium we had 600 attendees signed up.

The author welcomes the Dalai Lama to Erasmus University in Rotterdam, 2014.

8
Fourth dialogue: Education of the Heart (Rotterdam, 2014)

The fourth public dialogue generally followed the format of the third dialogue in The Hague, but this time we had students of all ages participating in the discussion. The overall theme was to explore what type of education is required in order to create a sustainable economy. We asked the Dalai Lama to reflect on how to develop "the heart" of people, not only in schools but in our communities and companies as well. The symposium started with a lecture from the psychiatrist Dan Siegel on the functioning of our brain and heart. The Dalai Lama then went on to emphasize the following points:

- The 21st century requires a different type of thinking from that of the previous century. Since the world is interconnected, we should think in terms of the whole of humanity. We now need genuine cooperation among 7 billion human beings—that is the only way forward to solve the world's challenges.

- There needs to be awareness of a new interdependent reality and, on the basis of that awareness, we can develop new mental models, expanding the "in group" and shrinking the "out group." Since many people are not religious, we need "secular ethics" to emphasize the many benefits of compassion and cooperation.

- Compassion is not just a warm feeling of empathy, but a practice that involves critical thinking and the courage to act. A mind trained in compassion will have inner strength and self-confidence. This will help to overcome the stress and anxiety that is part of the modern world and is especially needed for those engaged in the leadership of social change.

- In education it is not enough to develop the cognitive side— there should be more emphasis on cultivating inner values, such as compassion and warm-heartedness. This is especially relevant in business education, as these qualities can guide leaders in dealing with an increasingly interconnected and transparent world.

In the morning of May 12, 2014, on a beautiful spring day at the campus of Erasmus University in Rotterdam, a sold-out crowd from education, including teachers and students, young children and adolescents, and people from the business world, came together to listen to the Dalai Lama and to one another. The conference presented a range of best practices and inspiring examples from the educational field in Holland, spanning education from primary school, secondary school, vocational education, and academics up to management education.

Ruud Lubbers: It is a joy to be together again with Your Holiness and to continue the dialogue that we started 15 years ago, this time focusing on education—how we can improve our education system to build more compassionate societies and sustainable economies.

8.1 What does the "education of the heart" mean?

Dan Siegel: The purpose of our meeting here is to really understand what is meant by the education of the heart, something that—as we agreed among ourselves this morning and in previous conferences with you—we need more of in the world, especially in the world of management and leadership.

PHOTO: SARAH WONG

Ruud Lubbers and the Dalai Lama, Rotterdam 2014.

In your book *Beyond Religion*,[1] you talk about secular ethics, and having values that can come from the inside without being formally attached to any particular religion. The religions may have a wonderful way of teaching values, but that for a 7 billion-strong human family to come together around human values we need a secular ethic.

We learned that educating the heart occurs in a number of ways. First is the idea that you have to empower the individual to look inwardly. This doesn't happen through a formal meditation practice necessarily, but by saying that you have a passion or feeling of inspiration inside of you. This feeling isn't something that we would measure in a test, or something that we would put in a textbook, but something that you can express through a relationship. So some of these empowerment techniques take place through music, or in a classroom, or on the job in companies. They all come down to a process of getting in touch with a personal passion.

The second point that emerged was how altruism can become a natural part of reaching out to others—to be compassionate, to feel people's suffering and actually want to reach out and help. And finally, the third point was that somehow there needs to be a way of stepping outside of the usual

1 Dalai Lama, & Norman, A. (2011). *Beyond Religion: Ethics for a Whole World.* New York, NY: Houghton Mifflin Harcourt Publishing.

learning processes and perceiving what is really going on, and from there realizing that you can stand up to the way the current educational system works.

This morning we heard representatives from the educational system, from consulting firms for the government, from companies who look at what competences a future manager needs, and professors of business education. What became clear in the discussion was that the system of education can't be changed easily. It is part of a structure of institutions, laws, budgets, and controls that has formed over many decades.

My first question to you, therefore, is: what advice do you have to all those who have the courage to create more education of the heart and to develop a kind of internal set of values for educational innovation? What have you found in your travels and dialogues with many experts that we could learn from?

The Dalai Lama: First, I believe the subject of wholeheartedness is a very important one. Our future world needs wholeheartedness in all layers of society and particularly in business. Second, I consider it important that many young school and university students are here. When I meet young people, I also feel much younger. I think that the generation of Mr. Lubbers and myself, the generation of the 20th century and also the generations of our ancestors, really created a lot of trouble and a lot of violence.

However, the 20th century has now gone. This is a matter of fact: the past has gone and nobody can change the past. The reason why we should look back is that we can learn something from past experiences in order to create changes for the better. The future is totally open, like empty space, like a feather blowing in the wind—anything is possible. The 21st century is just beginning. It is up to all these young people, younger than me, to make a better and happier world. The future is in their hands. So please don't follow the 20th-century generation who, whenever some sort of disagreement arose, first thought in terms of solving the disagreement by force.

That type of thinking is outdated. Now that the world has become interconnected, we should think in terms of our entire humanity, "my world" rather than just "my nation." The modern reality of global problems, such as global warming and the global economy, brings us more closely together. We now need genuine cooperation on the basis of

> That type of thinking is outdated. Now that the world has become interconnected, we should think in terms of our entire humanity, "my world" rather than just "my nation."

a sense of oneness of 7 billion human beings. Whether you like it or not, this is the only way forward. For example, here in the Netherlands it is quite clear that your country's future depends on Europe, the EU. While it is also clear that the future of the EU depends on the rest of the world. So we now live in a time that East needs West, West needs East, South needs North, and North needs South.

Even though that is the current reality, much of our thinking is still strongly dominated by "us" versus "them." That creates a gap between reality and our perception. Any effort that is based on a distorted sense of reality will create a disaster. (Even if you want to hurt someone, you first have to make an accurate assessment of reality; otherwise you will not succeed in your goal of harming that person.) So with a view on the well-being of all human beings, we now need realistic action, that is, action based on clear understanding and a clear vision.

Therefore, I am very happy to see these young people: you are very important in determining the most realistic action for the future. This was the experience of the Tibetan refugees when we had to leave our homeland. We gave our young people every opportunity to learn and study, first in India where we settled, later also in the West with the help of international aid organizations. Through the support that we gave them and the knowledge that they acquired, the Tibetan community in exile survived. It was a collective effort from the side of the entire Tibetan refugee community. This shows that realistic action, combined with mutual affinity and support, can create success.

When I listen to the news (I listen to the BBC on the radio every day) I always hear of murder, corruption, violence, cheating, and starvation. And on my travels to many different places, including Africa and Latin America, I have seen firsthand the huge gap between the rich and the poor. Often the gap is growing, as the rich don't take care of the poor. In religion too you find people who are self-centered, corrupt, and violent. As in politics, where we use the word "dirty politics," perhaps these days we can also speak of "dirty religions." It is interesting to reflect on this point. Politics in itself is not dirty—it is just an instrument for serving people—but many of those who are involved in politics are not very clean.

My point is: in any human activity, if the motivation and moral principles are lacking, the outcome becomes dirty. If the people involved lack moral principles, whether it is in the economy, education, politics, or religion, whatever the professional field, the outcome is failure. When people have a strong self-centered attitude, their morals will be weak and they will be tempted to cheat, bully, steal, or even kill. In this way, even religion can become destructive.

So, when we observe our humanity, we can conclude that we need some effort to bring moral principles into humanity's mind. Without this effort, self-centeredness will remain. Now the question is: how can we bring these moral principles to 7 billion human beings? Traditionally, moral principles and human values were promoted through religion. But we have many different religious traditions. On top of that, out of all human beings, over 1 billion are nonreligious. This means that we need to find a secular way to educate people about values and warm-heartedness.

> when we observe our humanity, we can conclude that we need some effort to bring moral principles into humanity's mind.

On a biological level, we all know what compassion and warm-heartedness is and how important it is for our survival. It arises spontaneously and instinctively in mothers for their children, for example. But this type of compassion is limited: it is usually oriented to a small group of relatives and does not extend to strangers and enemies. Nonetheless, we can take this seed of compassion and use our human intelligence to develop this further. This is definitely possible. And here the role of education comes in.

Let me give an example. If you have a sense of warm-heartedness, you will have many more friends, and among them some trusted friends. Perhaps among your fellow students, you may find someone who is always cheating or bullying, perhaps it is someone with a rich family background. You may dislike him, but if you look more closely, you will find that this student may not have many genuinely trusted friendships. This shows that warm-heartedness is more important for trusted friendships than status or money.

> On a biological level, we all know what compassion and warm-heartedness is and how important it is for our survival.

Another example: if you see a beautiful girl or boy, on an external level you see beauty, which leads to attraction. But when you spend time with that beautiful girl or boy, gradually other things become more important. If you live with that person over many months or years, you may end up fighting. Inner beauty is more important than outer beauty. Inner qualities such as honesty, not telling lies, warm-heartedness, and respect are needed for long-lasting friendships and relationships. This is a fact, isn't it?

We human beings are social animals; at the most primitive level we need friends. Also, we need health—physical and mental health. Without these we cannot be happy. Now, this cannot be obtained by money, power, or other external conditions. Instead, all of this depends on warm-heartedness. Medical scientists have discovered that if you suffer from constant fear, anxiety, anger, and hatred, you gradually undermine your immune system. If you feel warm-heartedness, you will experience inner strength, which in turn reduces anxiety and stress. That is a very important factor for good health.

> If you feel warm-heartedness, you will experience inner strength, which in turn reduces anxiety and stress. That is a very important factor for good health.

I am not talking about next lives, heaven or hell, but about how we can create a happy human family for all people. I believe we need secular education that is suitable for believers and nonbelievers, that helps people to develop warm-heartedness in daily life. This should be promoted in a practical way, backed up by science, so that people can gain experience and from there develop their own convictions. Don't rely merely on religious references such as "Jesus Christ said this, and Buddha said that," because many people don't care.

You have to make spirituality relevant to day-to-day life, only then will people pay attention. We are very aware of the need for physical health, now we need to be more aware of the importance of emotional health. There is much knowledge these days on the workings of emotions and how they lead to well-being that can be applied in education.

Dan Siegel: Let me lay out a couple of scientific things that we believe are true, and ask for your suggestions on how education of the heart might build on these points. The first thing to say is that our mental experience of emotions or thoughts, or even the way we make decisions, seems to be related to, not the same as, but related to what goes on in our body as well as our brain. One thing we know about the brain is that, over the period of evolution, human beings have evolved and survived because they seem to have made the distinction between who is in the "in group" that they can trust, and who is in the "out group" that they should be wary of and avoid. From an evolutionary point of view, knowing who is in the "in group" versus the "out group" has become an important function of the brain. Another factor is that, in research studies of people under threat, the threat increases the brain's tendency to make a distinction between the "in group" and "out

PHOTO: SARAH WONG

The Dalai Lama speaking at the "Education of the Heart" dialogue, Rotterdam 2014.

group." So, given the huge numbers of human beings, given the challenges we face with depleting natural resources, economic recession, and ethnic and religious violence, we seem to have many reasons for feeling threatened.

I believe you are saying, and this would change my view as a scientist, that the love you receive in your family, such as the love you get from your mother, helps people move beyond their brain's vulnerability to make the in/out group distinction and that we are in fact all a part of the same "in group."

I would say that the "out group" really reflects our sense of fear and violence. If we consider carefully what is really threatening us, it is the fact that we are not seeing how interdependent we all are. People must move beyond the emotions of fear and withdrawal and into a state of openness.

But how can we help people to develop that experience of an interconnected, expanded, distributed self?

The Dalai Lama: I think your question relates to something we have seen in the development of human civilization, where smaller groups started to collaborate with other groups and ultimately created some sort of integration for the sake of their own survival. For example, let's look at the European Union. I don't think that, in the early part of the 20th century, people on this continent were thinking of the oneness of Europe. Now many European nations are very much aware of their common interests, which drive a process of integration. So it seems that through awareness of a new reality after the Second World War, the concept of the EU emerged. This does not come from merely instinct, right? Rather, the EU arose from being aware of the new reality of interdependence and that collaboration was the best way of serving common interests.

The interesting point is that now there is almost no danger of violence or war among the members of the EU. In contrast, if you look at Africa, there are many small nations fighting each other. The same applies to Asia, where now India and China don't see themselves as members of the same group, and to Russia, a country that seems isolated from the West. As long as these nations feel isolated, there will be mistrust among them, which causes fear and anxiety, and may turn into violence. So, in short, there needs to be awareness of a new interdependent reality and, on the basis of that awareness, we can develop new mental models, expanding the "in group" and shrinking the "out group."

> There needs to be awareness of a new interdependent reality and, on the basis of that awareness, we can develop new mental models, expanding the "in group" and shrinking the "out group."

Of course, we need to overcome certain fears. I make a distinction between positive and negative fear. Fear is sometimes necessary, and even anger can sometimes be necessary, when you need energy to defend yourself.

But often fear and anger arise not because of a real external threat, but out of some sort of unhappy, anxious, or agitated state of mind. When you have this unhappy state of mind and a small problem arises, you react strongly. In many cases, you overreact in fear and anger. This type of emotion is unnecessary. Why do we overreact in this situation? Because deep inside ourselves, we feel weak and powerless, we lack self-confidence, and so we blame external conditions.

In contrast to this, a state of mind with compassion brings inner strength and self-confidence, which in turn helps us to go beyond fear. If your mind is very self-centered and insecure, you will develop a sense of distance between "me" and "them," between the "in group" and the "out group." When this happens, we try to hide certain fears and we cannot carry out our life transparently. The result is more suspicion, more distrust.

On the other hand, if you have an open and compassionate attitude, in most cases the response from the other side will also be positive. This does not necessarily mean that you have to give in to the other side, that you should become weak and vulnerable. It is quite the opposite. If, in spite of your open heart, their response is still negative, then you have the freedom to act accordingly, possibly in a strong way. In this sense, you act out of free choice, not out of anger or fear. This is the realistic way of dealing with fear.

> If your mind is very self-centered and insecure, you will develop a sense of distance between "me" and "them," between the "in group" and the "out group." When this happens, we try to hide certain fears and we cannot carry out our life transparently. The result is more suspicion, more distrust.

8.2 Developing the "heart" in education and business

Luuk Stevens: I have made it my life's work to research and improve the quality of education. I came to the conclusion that the key to high-quality education lies in the relationship between teacher and student. In modern education we are too preoccupied by issues that I believe are secondary to this primary relationship between the student and teacher, which is where the actual learning is taking place. If there is no relationship between them, there will be hardly any learning that is taking place. I strongly support the ideal of education of the heart, but it will only have impact if we can make it practical for teachers in the classroom—for they hold the key to the heart of the student. Could you please suggest one or two best practices of teaching that would make the difference in the classroom? Is there something that a teacher can do?

The Dalai Lama: The best way to teach warm-heartedness is to demonstrate it through your own action. Teach students as if they are your own family member and take full responsibility for their whole life. With that kind of attitude, while you teach a certain subject, you take full responsibility for their future. In this way, the relationship between teacher and student becomes a genuinely close human relationship. The result of this is that the subject that you are teaching goes deeply into the student's mind, much deeper than if such an attitude was lacking. Conversely, if students in the classroom don't show any affection to the teacher, then the class can become a burden to the teacher. So warm-heartedness has to come through your own actions. On top of that, according to Buddhist literature, the teacher should be aware of the

> The best way to teach warm-heartedness is to demonstrate it through your own action.

different mental dispositions of the students. Depending on the student's mental disposition, the teacher should teach in different ways. It starts with your attitude but you also need to have certain teaching competences.

Rob van Tulder: As an educator, I use the following mantra to inspire my students and myself: "You have to use the warmth of your heart to keep your head cool and your hands productive." I recognize this in much that you say. My question is as follows: today we are at a university known for its teaching on economics and business, where we train the managers and leaders of the future. My focus is on the role of business in society and how we can ensure that companies add value to society and not detract value from society or nature. You spoke many wise words on education in general, but can you perhaps reflect on the role of business in society?

The Dalai Lama: The teaching of values, such as warm-heartedness and care, is even more important for managers and businesspeople! Through their organizations, products, and services, they have responsibility for many people. Especially today, with information technology, managers can no longer hide away from the public and the workers. Everything is becoming transparent. Previously, it was difficult to see what business leaders were doing behind closed doors, what were their decisions and what was the impact from their decisions. This is now changing. If managers have a sense of responsibility for their actions, they will be more inclined to consider the far-reaching consequences of their actions. They will not just follow their own interests, but also those of their employees, clients, and so on, including future generations as they can make decisions with a longer-term view than just this present quarter or this year. I believe that these types of decision will create more benefit for all those involved, including business benefits! There should be no contradiction between longer-term benefits for society and short-term financial benefits for an organization.

Leaders with this attitude will be better equipped in today's transparent world than those who are not. They have nothing to hide. I think this is the best way to ensure business ethics and accountability. As this approach comes from the manager and not from an external rule or regulation, it will be more effective. If people have the wrong attitude they can always circumvent rules; no matter how strict the rules are, the manager will find a loophole. I think that business schools have a particular responsibility here in teaching managers to develop these values.

A student: As an individual, I can aspire to be happy. But on the macro level of society and economics this is much more difficult, especially since business and investors seem to be ruled by a selfish gene. How can we overcome this?

The Dalai Lama: I am not against selfishness, that is, the impulse to take care of yourself. Self-care and self-compassion are good. In the West some people seem to suffer from self-hatred—this is totally useless. But there is wise selfishness and foolish selfishness. If you pursue your own interest without care or consideration for others' interests, then this will become self-destructive. This will become the seed of the end of your business or investment because, as a business, you are dependent on others.

> I am not against selfishness, that is, the impulse to take care of yourself. Self-care and self-compassion are good.

Modern science says there is no selfish gene. They looked in the brain but they cannot find such a thing. What they did find was the innate appreciation of kindness and compassion, because we all depend on that. When we were born we were totally dependent on the kindness of our mother. This applies also to students, teachers, and businesspeople!

A student: In our studies on business and management, there is often a belief that money is the same as success. We assume that when companies and people generate a lot of money, then they are successful. This assumption is rarely questioned in business education. But when you look at rich people, they usually don't seem so happy. In your mind, why are they so unhappy?

> Money cannot deliver inner peace. Only kindness brings inner peace.

The Dalai Lama: This is a clear indication that money cannot deliver inner peace. Only kindness brings inner peace. For example, if you are very lonely and miserable, money cannot help you. Money does not have the capacity

PHOTO: SARAH WONG

The Dalai Lama is welcomed by the youngest participant at the "Education of the Heart" dialogue, Rotterdam 2014.

to give you affection. If you feel miserable, maybe your dog can show some sort of affection, by licking you. But money does not have that capacity.

A student: What do you remember as the most inspiring teaching or subject during your own childhood education?

The Dalai Lama: When I was at a young age, I found education really boring and not of much interest. I was only interested in the next holiday. I think you also have that experience, right? Then, gradually, by growing up and being confronted with difficult circumstances, I started to see that certain knowledge can really help you to maintain peace of mind when dealing with problems, in my case major political problems in Tibet. What I found most useful is Buddhist psychology, in addition to the Buddhist logic. The Buddhist view of reality helps you to look at problems from a broader, more holistic, perspective. You will come to see, for example, that difficult things happen due to many different causes and conditions, not simply one cause. This will diminish the intensity of your worry or your anger. Scientists have observed that when we develop anger, then in our mind we need some kind of independent object, something we can isolate and determine as the real troublemaker. But when you learn to look at problems in a more holistic way, you can't find that independent or isolated object. We only find an interdependent reality, in which everything is dependent on other factors. So this type of holistic education helped me when facing many difficult circumstances; through establishing peace of mind I could effectively deal with problems.

8.3 Applying the education of the heart in practice

A student: I am growing up at this great school with free space to learn, to follow my heart and discover my talents. I can learn anything I need for my life in the future, but what do you think the future needs from my friends and me? The future is very uncertain so it is not clear to us what knowledge and skills will be needed in the future. What is your advice?

The Dalai Lama: I don't know! It depends on your own particular predisposition and your personal interest. It is difficult to say. You should think and explore more.

But I have one observation to share. The existing way of life, in the West, and now all over world, is very much oriented toward materialistic values. This is the case in business, but also at government levels and family levels—there is so much talk of money. Whole modern society seems to revolve around material values. I think this also exists in the modern education system. As I said before, when we are very young we receive a lot of affection and we mostly play; our values are around caring and sharing. But when we join a school, we are slowly exposed to material values. We learn to compete in education to get high grades, even in music and sports. We learn knowledge and skills that will give us good jobs with good salaries.

So, naturally, as a child in that environment you automatically take on these material values. But is this good for the child and his or her role in society? I feel that education should put more attention on ethics and warm-heartedness, on inner human values. I am aware that a number of educational scientists in America and India are seriously carrying out research on developing a curriculum on secular education based on the connectedness to the larger human family. I think

> I feel that education should put more attention on ethics and warm-heartedness, on inner human values.

we should learn from what they discover and conduct further experiments. If the effects are good, we can extend it from one school, to ten schools, a hundred schools, and so on.

A teacher: At the Avans University for Applied Science, we have a mission to show students what their talents are and how to use them, as well as the knowledge that we provide through our teachers. We felt that there was something missing and so we created a program called "Avans Flow," where students can discover and develop their inner talents in different ways. We use mainly mindfulness meditation, creative arts, theater, and lectures on reflective subjects. In all our programs, we start with a guided meditation for one minute, to connect our hearts with the soul and the hands. We begin with inwardly connecting to the Earth, the Earth we all share, and then we go to our heart, and connect with our soul and mind. Can you give us advice on the best form of meditation? What do you think of the style of meditation that we have developed?

The Dalai Lama: There are basically two kinds of meditation in the Buddhist tradition. One is analytical meditation and the other is single-pointed meditation. From my own experience, analytical meditation is more effective. Single-pointed meditation is usually done after analyzing certain things, for example compassion. Think of the value of compassion, and think of the destructiveness of anger and hatred, which is the opposite of compassion. Then you develop some kind of conviction that compassion is really wonderful. Then, think about the basic oneness of all human beings. You can even extend that feeling of oneness to insects, birds, and all animals. Reflect on how much these innocent sentient beings suffer. After experiencing a feeling of concern for these beings, you develop the wish, "how nice it would if their suffering disappears." When you think like that, develop some kind of enthusiasm.

After the analytical part of the meditation, you can move to the single-pointed part, which is meditating without investigation. Place your mind completely on the experience of compassion for other sentient beings. You can take any topic of meditation but of course a positive emotion helps to make you a better, happier human being. I think that most of you meditate on money, right? You concentrate on your salary or profit. And you calculate your budget: if I use daily this much, then how much is left? These are also analytical meditations! But does it make you happy?

A teacher: Can you please teach us some meditation technique that we can do at our school?

The Dalai Lama: I would like to share one technique to calm down our mind. [The Dalai Lama shows the nine-breath movement meditation.] This certainly neutralizes your mood, your emotion. When you have an agitated mind, just concentrate on your breathing and drop the feeling of agitation. Then your mind becomes comparatively calm. Now, meditate on love and compassion. If you believe in God, think of his infinite love. We are the creation, we are one with the creator. We all have a spark of God, and that is infinite love. If you don't believe in God or any other religion, you can simply think about basic human nature, which is generous and compassionate. Now concentrate for one minute on love and compassion.

Mark Simons: I represent Mountain Child Care, which is a social enterprise that helps young people from both Nepal and Holland with their personal development by discovering their life purpose and integrating that with their work and career. This includes young professionals who have just started work in companies. We do that by taking them on a trek, both in the Himalayas and in the Netherlands. In Nepal, we walk for six days in the high mountains and it is physically and mentally very tough. During the trek we do all kinds of exercise, including yoga, meditation, visualizations, coaching, and dialogues. The results are exceptional. My question is: what would you suggest we could we do better so that many people open their hearts and discover their life's mission?

The Dalai Lama: I don't know. You have the practical experience and, if this experience is beneficial, this should be continued! That would be wonderful. Perhaps you can add one thing. Every morning, when you wake up, you should form an intention, such as, "Now this day I will devote myself to the service of other people. In this way I will make my life meaningful." That is important addition to whatever work you are doing. Your motivation is not for money, not for fame, but only for serving other human brothers and sisters who need help and education.

If you do this there will be some benefit for yourself too. Once you have set this motivation, then even though you may feel physically exhausted at the end of the day, mentally you will still feel fresh. This is my own experience. Each morning, I dedicate my body, speech, and mind to the well-being of others, and when I look back in the evening I may feel a little bit exhausted but I also have a sense of fulfillment of having served a purpose. I think this will be the same for you and your participants.

In fact, I think this type of discussion makes us human. When I meet people, when I talk with people, I always think, "here is another human being." On a primary level—mentally, emotionally, and physically—we are the same. You want a happy life. I also want a happy life. Sometimes I have an emotion that creates some problems, you also occasionally have an emotion that is disturbed, don't you? We are the same! So, on that level, we can communicate very easily. If I focus on the fact that I am the Dalai Lama, I am the holy Dalai Lama, then I have to keep up a very holy persona all the time. This would be self-deception. If we identify with our differences, then there will be more divisions and more distance between us. This will result in feelings of loneliness. If we extend our hand as human brothers and sisters, we can feel very happy deep inside. But if I place too much emphasis on the thought that "I am the Dalai Lama," then there cannot be many brothers or sisters. If I think that I am a human being, then suddenly I have a lot of human brothers and sisters. Is this clear? What do the students think? What do you feel?

A student: I agree with you that in this world we are all brothers and sisters and we all have the same fears and joys.

The Dalai Lama: Very good, thank you.

8.4 Dealing with stress, grief and fatigue

Mark Woerde: I am a social entrepreneur and I have a question about an issue that I have been working on, which is a film about online child abuse. This is becoming a major problem through the internet and the income gap between rich and poor countries. It is also a very sensitive topic and difficult to solve, so many people close their eyes to it. When I studied this problem and saw the suffering of innocent young children, I felt a lot of pain and stress. My question is: how can we deal with the stress that comes with working on serious social issues? What advice do you have for the social entrepreneur or social worker who is experiencing the pain of the suffering of those children? I think this is relevant to all social entrepreneurs who have identified a serious social problem to work on. How can we keep ourselves well in the face of carrying out our important work in reducing the suffering of others?

PHOTO: SARAH WONG

From left to right: Ruud Lubbers, the Dalai Lama, his translator, and Daniel Siegel.

The Dalai Lama: Through the practice of compassion, which involves a number of steps. *First*, when you develop a sense of compassion for another sentient being, you will first share in their feelings of suffering. This is good, otherwise you would remain indifferent and indifference does not lead to action, only inaction. On the basis of this shared sense of suffering you can develop genuine compassion, which is a sense of concern for the other. However, if you leave it at that and only experience the feeling of suffering, you yourself may become demoralized and you may develop a state of hopelessness. This is not good, as it would lead to inaction. So in the practice of compassion there is an important *second* step. In addition to being concerned, you should use your intelligence. This requires you to keep some distance from the problem that you observe and analyze. What are the causes and conditions? What can you do to resolve the situation?

Dan Siegel: There is a study in which photographs of a car accident are presented to two separate groups and their responses in the brain are monitored through a brain scan. One group of people is asked to imagine if that were you in the car crash. The other group is asked to imagine what it would be like to be that person. In other words, in the first group there is a sense of identification, and in the second there is not, which is a small but significant difference. The first group became overwhelmed by their emotions and the neural circuitry of their compassion, which includes neural networks right behind our forehead, shut down. The second group, on the other hand, actually maintained emotional balance. They could activate the circuits of compassion that allowed them to not only empathize with the other person's feelings, but to actually get ready to take action and to support them. This is a very important study that illustrates exactly what you're saying.

So we need to have a kind of healthy distance and to maintain an analytical mind in a sense, just to keep our own balance.

I remember you told me a few years ago that it isn't just in spite of the suffering that we need to be joyful and have a sense of humor and be playful, but it's actually because of it. If we let our playfulness and our humor and our joy for life be smothered by the suffering of the world, then suffering will have won. That was very helpful for me to hear those words.

The Dalai Lama: That is very true. Actually, genuine compassion comes through training in a step-by-step manner. One important aspect of training in compassion that I did not mention is courage. This is in fact the *third step* after first experiencing a shared sense of suffering, or empathy, and, second, using your analytical mind to observe and analyze the situation well. The third step of courage is needed to develop the will to help those who suffer. Feeling empathy is not enough; you need to analyze as well and then develop courage and self-confidence in order for compassion to be realized. It is my experience that someone who has developed compassion to this level will not despair. Through his analysis and courage he will find a way to act appropriately. He cannot feel overwhelmed because he can use the problem as a means to develop his compassion and determine a way of action.

> Feeling empathy is not enough; you need to analyze as well and then develop courage and self-confidence in order for compassion to be realized.

Dan Siegel: There is a professional term called "compassion fatigue." Some people suggest we should change that term to "empathy fatigue,"[2] because in your teachings and other research, when the first step is empathy, we are experiencing another person's feelings but we have not yet developed compassion as you describe it. Compassion comes from the additional steps of

2   Stebnicki, M.A. (2007). *Empathy Fatigue: Healing the Mind, Body, and Spirit of Professional Counselors.* New York, NY: Springer.

deciding how we can help to reduce the suffering and mustering courage for that, which research has shown touches a deep source of well-being inside ourselves. The empathy fatigue will not arise in this case.

The Dalai Lama: Yes. In dealing with problems—any problem, including these business problems—or tragedies such as major social issues you need to have a degree of courage and self-confidence. Shantideva, a great Buddhist philosopher, said: "When we face a problem, analyze the nature of the problem. If you find that the problem can be overcome, then there is no need to worry."[3] Make the right effort to overcome it, and then the problem should lead to increased courage and self-confidence. In this way, the problem can become the source of your strength. On the other hand, if the problem cannot be overcome, then there is no use worrying. So in both cases, there is no cause for worry. This is the best approach to problems or tragedies of any kind.

Dan Siegel: I would like to build on the incredible importance of that wisdom, especially for the young people here. We are facing major economic and ecological challenges, such as unemployment and climate change, which are very upsetting. They can cause us to feel despair, especially when the reports keep on coming out that it is worse than we thought. How do we take care of our minds and our relationships with one other? Can we stay actually awake to difficult issues such as climate change and try to do something about it, or do we allow the feeling of despair to overwhelm us and then withdraw and don't do anything about it?

A student: I still find it difficult to think about how to apply this thinking. Can we learn from you how you deal with things such as criticism, loss, and grief?

The Dalai Lama: I do the following things. First, I use common sense—I try to put things in the right perspective. For example, you can reflect on the fact that your problem is small in comparison to many of the other problems that exist in the world. Second, you can also think along the lines that if you are criticized, then at least you are not hit. If you are hit, then at least you are not killed. If you are killed, then that is the end and so there is also no need to worry. You also need to analyze the situation carefully: if I am

3   Shantideva (1997). *A Guide to the Bodhisattva Way of Life* (V.A. Wallace, & B.A. Wallace, Trans.). Ithaca, NY: Snow Lion.

criticized, what are the grounds? Maybe the one who criticizes me is right, in which case it is good that he points out my mistakes. If not, if there are no real grounds for the criticism, then I can develop a sense of pity, a sense of compassion for the one who criticized me wrongly. Another method, from Buddhism, is to reflect on the ultimate nature of things. This is similar to the insights from quantum physics, which show that there is nothing that is permanent or independent. Everything exists as a temporary phenomenon, in dependence of causes and conditions, including the way you look at it. Your perception determines a large part of how you experience the situation. According to the research of my friend, the psychologist Paul Ekman, when you are angry or full of desire, your perception changes the object by some 90%. This means that only 10% of your perception is correct, and the rest is imagined.

> Everything exists as a temporary phenomenon, in dependence of causes and conditions, including the way you look at it. Your perception determines a large part of how you experience the situation.

Sander Tideman: Thank you Your Holiness for joining us today. It has been more than 30 years since I first met you, when I was a young student like many here today, and I am glad that through this dialogue you have been able to inspire many young people like you inspired me when I was young. It feels as if we have closed a circle here—yet the inspiration continues. On behalf of the organizers and all the participants here, I would like to thank you from the depth of our hearts for your inspiration and your wisdom.

The Dalai Lama: I remember our first meeting. We all have the potential to create a happy atmosphere, to create a happy human family. So the changes that are needed to bring about a more compassionate society are not coming from the government or companies, or from the UN or the EU, but from individuals. You should create a compassionate atmosphere, first for oneself, then for your own family, and then, through that one family, you can influence a hundred families, and ultimately we will create a more happy and peaceful society. So, irrespective of whether you believe in any religion or ideology, each one of us has the potential to bring positive change. I am very encouraged that in the meeting today we have made serious efforts to develop more compassion in management and education.

From left to right: the Dalai Lama, Daniel Siegel, Rob van Tulder, and the author, Rotterdam 2014.

9
Shared purpose: the case for societal leadership in business

It has been more than 30 years since the first meeting with the Dalai Lama sparked a quest for meaning in my life and in business. My journey began as a young lawyer with an interest in Asia. I then became a banker in China and other emerging markets when financial capitalism reached its zenith. The ongoing meetings with the Dalai Lama helped me to trust my intuition that something was wrong with the financial system—as indeed we saw during the 2008 financial crash and the collapse of seemingly strong and solid financial institutions and whole markets. In my mind, it marked the end of an era in the history of capitalism, as we had known it.

Inspired by my continued exchanges with the Dalai Lama, as well as my work with innovative leaders and entrepreneurs in the East and West, I reinvented my career at various times. This experience prepared me for what I now consider to be my true purpose: a mapmaker and guide to help leaders transform their organizations into a positive force for good in society and for the planet. This map starts with a framework for the future of business leadership, which I call the "societal leadership of business." This integrates what I have learned in (business) life and from the dialogues with the Dalai Lama.

The dialogues have resulted in six emerging themes that are highly relevant for business leaders who want to be a positive force for society: holistic thinking, compassion, responsibility, creativity, collectiveness, and education of the mind/heart. This chapter analyzes these themes and explores related factors in the field of economics, business, and leadership, bringing them together into the so-called "6C" model for societal leadership, which can be practiced by leaders. This model is intended to discover "shared purpose" by serving the needs of the business, the needs of customers, and the needs of the communities of people and wider society of which they are a part.

By training and expanding their worldviews, mind-set and attitudes, leaders can develop their understanding of levels of complexity (simplified into "I," "Org," and "All"), along with their capacity for compassion, which consists of self-compassion, compassion for others, and compassion for all. The result is a strong sense of "shared purpose" that empowers leaders into positive action. In this way, they can become societal leaders who are capable of delivering value for themselves, their customers, and society—in other words, creating triple value.

9.1 The changing context of business: new ways of thinking[1]

> The current crisis [in the economy, society, and environment] could teach us to look beyond material values and unrealistic expectations of limitless growth. When things go seriously wrong [such as in the financial crisis], it is often because a new reality is still being viewed with outdated concepts, and this is certainly the case [in the domain of] the economy today.
>
> The Dalai Lama

As this quotation from the Dalai Lama indicates, the changing operating reality for business calls for a fundamentally different way of thinking and seeing. This can be described as new worldviews or mind-sets that business leaders need to employ in order to understand and access modern reality for the benefit of their business and society. The philosopher Thomas Kuhn[2] defined these as a paradigm shift, meaning a fundamental change

---

1   The first section of this chapter draws on Tideman, S.G., Arts, M.C.L., & Zandee, P.D. (2013). Sustainable leadership: towards a workable definition. *Journal of Corporate Citizenship*, 49, 17-33.

2   Kuhn, T.S. (1962). *The Structure of Scientific Revolutions*. Chicago, IL: University of Chicago Press.

in the mode of perception, frames of reference, and underlying beliefs and assumptions. The French novelist Marcel Proust said: "The real act of discovery consists not in finding new land but in seeing with new eyes."[3]

This chapter explores how these new paradigms manifest in society and business at the levels that were identified during the dialogues. First, there is the "All" level, including the larger macro system and the principles behind it (rooted in economic and social theory). Next is the "Org" level, with organizational dynamics, business models, strategy, and performance. Finally, the "I" level represents the individual dimension, including leadership. At this point we will specifically uncover the new leadership mind-sets that are required to deliver performance for both business and society, how they can be applied to value creation at the "I," "Org," and "All" levels, and how they can be developed and trained. This will be described in an integrated model for societal leadership in business that could support the creation of a new generation of business leaders.

9.1.1 The economic system

The consensus that has emerged from the 15 years of dialogues is that global problems have been created (and persist) because political and economic leadership employs flawed and increasingly traditional economic and business systems, based on limited assumptions about the interdependence of economic, social, and ecological reality and the multifaceted drivers of human behavior.

These assumptions were mainly derived from Newtonian physics and Darwinian biology. This worldview, in which economy, society, environment, and wildlife were seen as separate worlds over which humans—the "fittest" among competing species—hold dominion. All this was done with the purpose of extracting value, against the lowest possible cost, and utilizing it for our human agendas (or to liquidate it to maximize GDP or quarterly profit margins). In this worldview, individuals and companies regard themselves as autonomous, individual agents who make their own rational choices—the image of *Homo clausus* or *Homo economicus*—in a relatively static and predictable context. The economist Milton Friedman expressed

---

3   Proust, M. (1923). *Remembrance of Things Past*. Vol. 5: *The Captive* (C.K. Moncrief, Trans.). Retrieved from http://gutenberg.net.au/ebooks03/0300501.txt.

this worldview in the famous quotation: "the only business of business is business."[4]

This way of thinking was the cornerstone of the industrial age when both natural and human resources seemed abundant and inexpensive. Although its underlying worldview has since evolved through insights from sciences such as psychology, biology, sociology, and ecology, it is no longer fit for purpose. In fact, the rather simplistic ideology of economic activity is increasingly recognized as the prime driver behind the emerging "tragedy of the commons," in which producers, consumers, and financiers hold each other in a Prisoner's Dilemma: a race to the bottom of over-production/consumption/borrowing and consequential ecological overshoot and social inequality. Given the fact that we have finite common resources for a rapidly growing population, by continuing to focus primarily on our own short-term business interests, we collectively end up as losers.[5]

Fortunately, there is a new worldview emerging that is more suitable to the modern context. It is a view in which people, business, economy, environment, and society are no longer separate worlds that meet tangentially, but are a single, inseparable entity. As they are interconnected and interdependent, decisions need to be made by looking at the complete picture. This matches the view of the sociologist Norbert Elias, who said that humanity should see itself as *homines aperti*—people are in open connection with each other and their environment, being formed by and dependent on others and nature.[6] This view of interconnectedness has meanwhile been confirmed by findings from many scientific disciplines, most notably psychology and social neuroscience.[7] The financial crisis in 2008 made it clear that this interconnected worldview is not merely academic: it best describes the reality of global society, business, and finance, which functions as a tightly interwoven web of human relationships and interaction. In the new reality, "business as usual" is no longer an option from a long-term

4   Friedman, M. (1970, September 13). The social responsibility of business is to increase its profits. *New York Times*, p. SM17.

5   Gilding, P. (2011). *The Great Disruption: How the Climate Crisis Will Transform the Global Economy*. London: Bloomsbury.

6   Elias, N. (2000). *The Civilizing Process: Sociogenetic and Psychogenetic Investigations*. Oxford, UK: Basil Blackwell.

7   Seligman, M.E.P. (2002). *Authentic Happiness: Using the New Positive Psychology to Realize Your Potential for Lasting Fulfillment*. New York, NY: Free Press; Siegel, D. (2009). *Mindsight: The New Science of Personal Transformation*. New York, NY: Random House.

survival viewpoint. Indeed, leading companies have recognized the new reality—which is generally labeled as "sustainability"—as the next business "megatrend," just like IT, globalization, and the internet before it, determining their long-term viability as a business. In the words of the management scholar Frank Horwitz: "The only business of business is *sustainable business*."[8]

The shift toward sustainable business implies a departure from the linear three-pronged production–consumption–financing model in which money is abundantly made available by banks, to a more holistic life-based model in which constraints in financial, natural, and ecological resources are recognized as normal and consumers are recognized as citizens. It is a shift from the speculative debt/growth economy to the real economy, not only in a macroeconomic sense but also in terms of understanding the *real* drivers of economical value and sustainable business performance. The rules of the new economic game are no longer only to maximize return on invested capital, but also to create optimum resilience of the system by enhancing well-being, value creation, and performance of all the participants within the system.

In the 2009 dialogue, the Dalai Lama describes this shift as follows:

> The new reality is one in which the challenges facing humanity are "beyond individual effort" and our interdependences have become even starker. The gaps between our perception and the new reality are based on having concepts from the previous century in our minds, and this wrong perception creates the wrong approach. In business, these 20th-century concepts include management approaches that focus on maximizing short-term profits (not on delivering longer-term "triple bottom line" outcomes), ignoring activists or critical stakeholders, only caring about a regulatory license (not the broader "social license") and an "us and them" approach to problem-solving.

A similar sentiment is expressed by the leading business thinker, Gary Hamel:

> The biggest barrier to the transformation of capitalism cannot be found within the observable realm of org charts, strategic plans and quarterly reports, but rather *within the human mind itself* ... The true enemy of our times is a matrix of deeply held beliefs about what business is actually for, who it serves and how it creates value.[9]

8   Horwitz, F.M., & Grayson, D. (2010). Putting sustainability into practice in a business school. *Global Focus*, 4(2), 26-29.

9   Hamel, G., & Breen, B. (2007). *The Future of Management*. Cambridge, MA: Harvard Business School Publishing.

This paradigm shift has far-reaching consequences for leadership. On a macro level, leaders will be asked to create economic growth for a growing world population, while the natural resource base required for this growth will be increasingly insufficient. At a micro level, companies will have to come to terms with sustainability as a major business trend that requires drastic business model transformation.

Before exploring what this means for business leaders in particular, we will review some trends in economics and business thinking that form a part of the changing context, while pointing to aspects of a new, more appropriate worldview that could underpin efforts to create sustainable economic systems.

9.1.2 Mainstream economic ideology

> Modern science says there is no selfish gene. They looked in the brain but they cannot find such a thing. What they did find was the innate appreciation of kindness and compassion, because we all depend on that.
>
> The Dalai Lama

In order to explain the growing tension between mainstream economic theory and practice, the new field of behavioral and neuro-economics is increasingly called on to explain economic reality. This field has arisen over the last 30 years based on empirical findings from many experiments involving real people. It has achieved considerable traction thanks to the financial crisis, which made it obvious that more traditional notions of rationality and equilibrium of markets were mere theoretical constructs and had little to do with how markets behave in reality.

A whole range of recent publications has popularized these new insights.[10] A central insight of behavioral economics is that of fairness and trust as prime human drivers; neuro-economics (founded by Daniel Kahneman, who received the 2003 Nobel Prize in Economics for his studies on intuitive

---

10 Akerlof, G.A., & Schiller, R.J. (2008). *Animal Spirits: How Human Psychology Drives the Economy, and Why it Matters for Global Capitalism.* Princeton, NJ: Princeton University Press; Ariely, D. (2008). *Predictably Irrational: The Hidden Forces That Shape Our Decisions.* New York, NY: HarperCollins; Ariely, D. (2013). *The Honest Truth About Dishonesty: How We Lie to Everyone—Especially Ourselves.* New York, NY: Harper Perennial; Thaler, R.H., & Sunstein, C.R. (2008). *Nudge: Improving Decisions About Health, Wealth, and Happiness.* New Haven, CT: Yale University Press.

judgment and decision-making), explores the same territory of real behavior.[11] The significance of this work lies in its ability—for the first time in the history of economics—to describe the neuro-biological basis of economic behavior, thereby bridging the previously distinct disciplines of psychology and economics.[12]

This new neuroscience and behavioral science approach is revelatory because it provides empirical evidence derived from a biological basis for the notion that human nature is *not* driven by greed, materialism, extrinsic motivation, and egoism only; at least equally important are principles of cooperation, moral fairness, altruism, intrinsic motivation, and psychological well-being. This not only uproots the classical model of *Homo economicus*, but it also challenges the deeply felt belief that only external gratification through money and consumption can meet our needs. Over the last decades of economic *laissez-faire* policies, we have wrongly projected our happiness externally onto the idea of economic growth. This can be seen as the main cause for unsustainable ecological overconsumption. Creating sustainable economic systems, therefore, requires us to shift the objective of economic activity away from maximizing transactions for economic value, to balancing economical, social, and ecological value and well-being.

Today, there are plenty of examples that show the shortcomings in using outdated economic models. For example, while economic textbooks speak of money as a neutral mechanism that facilitates trade and investment, the way that most governments dealt with the aftermath of the 2008 financial crisis revealed the political nature of money and finance. Rather than merely facilitating the economy, it is now clear that money forms an inherent part of the debt-based economy, in which banks create money through interest-bearing loans as part of a political agenda to achieve vested economic objectives. As a result, when growth prospects disappear in times of crisis, banks turn out to be more entitled to money than taxpayers. This shatters the classical economic beliefs that money is a neutral instrument for trade, banks are normal private enterprises, and that governments in so-called free-trade economies adhere to free-market principles. The 2008 crisis has

11   Kahneman, D. (1979). Prospect theory: an analysis of decision under risk. *Econometrica*, 47, 263-291; Kahneman, D., Diener, E., & Schwarz, N. (Eds.) (1999). *Well-being: The Foundations of Hedonic Psychology*. New York, NY: Russell Sage Foundation.

12   Gowdy, J. (2009). *Economic Theory Old and New: A Student's Guide*. Palo Alto, CA: Stanford University Press.

shown that even the most powerful financial institutions and entire governments cannot escape the new reality of interconnectedness, dependence on human emotion, and dynamic market/social instability.

By extension, the mainstream view of markets as a neutral mechanism that efficiently processes our collective rational choices into collective well-being and a state of equilibrium has become obsolete. We now know, from both actual observations during the financial crisis and psychological research, that the minds of market players are continuously subjected to emotional and social influences.[13] Thus, the theory of market equilibrium needs to be replaced by a view of markets as dynamic, interdependent relationships that are both shaped by our mind-sets and choices and shaping our mind-sets and choices, mostly on an unconscious level.

Another flaw in mainstream economic thinking is the real price and profit of a product or business. There are many so-called externalities that are excluded or ignored in the process of determining price and profit, as anyone trained in accounting knows. An example of this is how governments measure the economic performance of their country. They apply the conventional indicator of gross national product, even though it is an incomplete measure because it excludes (and even undermines) critical resources such as water, clean air, and healthy people.

Finally, there are essential flaws in mainstream thinking about the value of nature. Classical economics assumes that nature and natural resources such as water, air, and land (unless privately owned) are free public goods. While the field of environmental economics has long sought to change this, national- and company-level accounting still do not acknowledge "natural assets" on the balance sheet. This may hopefully soon change as a result of the newest climate change agreement (which strengthens the practice of carbon emissions trading) and efforts spearheaded by a coalition of companies including the accounting industry to establish norms for natural capital accounting.[14]

Taken together, these shifts can be regarded as a transition from the **growth–transactional** model of economics, toward a more **sustainable–relational** economic model. The new economic theory goes beyond linear supply/demand exchange models to more sophisticated models of co-creation. It is no longer enough to know what we produce and consume

13 Zak, P.J. (2008). *Moral Markets: The Critical Role of Values in the Economy*. Princeton, NJ: Princeton University Press.

14 WBCSD (2012). *Measuring Impact Framework: A Guide for Business*. Geneva: World Business Council for Sustainable Development.

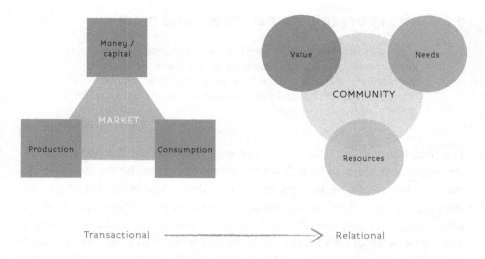

FIGURE 9.1   Shift in economic paradigm

and value this by the amount of money it has generated (the growth–transactional view). Of equal importance is understanding how to serve real human needs with renewable resources, and capturing the value created through this (the sustainable–relational view).

Matching real needs and resources entails a focus on the way we think and relate to each other. Given the central role of human thinking and interaction in the new economic paradigm, we should shift our perception of markets as anonymous transactional trading places to a community operating in an interdependent economic and ecological context. The members of the community are all interrelated stakeholders who are engaged in a continuous complex interdependent process of co-creation of value, while fulfilling needs, both short- and long-term. These needs go beyond merely material economic needs; they include emotional, social, and ecological needs. In parallel, the value created expands from (singular, financial) shareholder value, to (multiple) stakeholder value. This shift in economic paradigm is shown in Figure 9.1.

9.2 Shifts in organizational theory and practice

> When we search for organizations that have the capacity and ability to improve our world, global companies are at the top of the list. In particular, global corporations are in an ideal position to support developing countries in closing the gap between their economies and those of the most wealthy nations.
>
> The Dalai Lama

The field of organizational theory has seen a similar shift in paradigm. The transition at the macro level from the transactional to the relational worldview is also visible at the micro level of the business organization. The dominant image of an organization, as a hierarchically structured machine whose principal purpose is to make profit, is giving way to the metaphor of a living network in which value is generated by human relationships inside and outside the organization, based on a shared purpose. Contributions to this shift came, among others, from Edward Freeman who formulated the stakeholder theory,[15] the MIT scholar Peter Senge who introduced the field of organizational learning,[16] Jim Collins and Jeremy Porras who looked at firms that were "built to last"[17] and which were later defined as "from good to great" companies,[18] and Rajendra Sisodia and colleagues who defined companies who create profits by following purpose and passion as "firms of endearment."[19] This line of thinking corresponds quite closely to the views of the Dalai Lama, as evidenced by his reflection on the nature of business institutions:

> Companies are living, complex organisms and not profit machines. The profit should therefore not be the object of a company, but rather a result of good work. Just like a person can't survive for long without food and water, a company can't survive without profits. But just as we

---

15   Freeman, R.E. (1984). *Strategic Management: A Stakeholder Approach*. Boston, MA: Pitman.

16   Senge, P.M. (1990). *The Fifth Discipline: The Art and Practice of the Learning Organization*. New York, NY: Doubleday Currency.

17   Collins, J., & Porras, I. (1997). *Built to Last: Successful Habits of Visionary Companies*. New York, NY: HarperCollins.

18   Collins, J. (2001). *Good to Great: Why Some Companies Make the Leap. And Others Don't*. New York, NY: HarperBusiness.

19   Sisodia, R.S., Wolfe, D.B., & Sheth, J.N. (2007). *Firms of Endearment: How World-Class Companies Profit from Passion and Purpose*. New York, NY: Prentice Hall.

cannot reduce the purpose of a human to eating and drinking alone, we cannot regard companies solely as money-making entities.[20]

The initial calls for sustainable development came from concerned environmentalists and visionary social and public leaders, such as Rachel Carson, Donella Meadows, and Gro Harlem Brundtland,[21] typically projecting a vision of sustainability as a desired utopian "end-point," while putting the business community on the defensive as they were perceived to be the major threat to this desired state. As a result, companies responded with corporate social responsibility (CSR) in order to ensure a societal license to operate. In the years that followed, sustainability has been defined more as a change process in which business has gradually taken on a participatory and constructive role. The definition of sustainability gradually evolved, in the words of Peter Senge, from being "a problem to be solved, to a future to be created."[22]

When the bubble of financial capitalism burst in 2008, mainstream business leaders began to question the wisdom of the status quo.[23] It was no longer feasible for economies and business to "grow" themselves out of a crisis. With the growing environmental resource crisis in the background, the creative search for sustainable economic models has intensified. In almost any sector one can now find sustainability initiatives from leading businesses that Harvard scholar Bob Eccles has defined as "high sustainability companies" (HSCs).[24] Using a matched sample of 180 companies, Eccles and his team found that corporations that voluntarily adopted environmental

20   Eigendorf, J. (2009, July 16). Dalai Lama—"I am a supporter of globalization". *Die Welt.* Retrieved from http://www.welt.de/english-news/article4133061/Dalai-Lama-I-am-a-supporter-of-globalization.html.

21   Carson, R. (2002). *Silent Spring.* Boston, MA: Houghton Mifflin. (Original work published 1962); Meadows, D.H., Meadows, D.L., Randers, J., & Behrens, W.W. (1972). *The Limits to Growth: A Report for the Club of Rome.* New York, NY: Universe Books; World Commission on Environment and Development (1987). *Our Common Future.* New York, NY: United Nations.

22   Senge, P.M. (2008). *The Necessary Revolution: How Individuals and Organizations Are Working Together to Create a Sustainable World.* New York, NY: Broadway Books.

23   WBCSD (2011). *Collaboration, Innovation and Transformation: A Value Chain Approach.* Geneva: World Business Council for Sustainable Development.

24   Eccles, R.G., Ioannou, I., & Serafeim, G. (2014). The impact of corporate sustainability on organizational processes and performance. *Management Science*, 60(11), 2835–2857. Retrieved from https://dash.harvard.edu/handle/1/15788003.

and social policies many years ago exhibit fundamentally different characteristics from a matched sample of firms that adopted almost none of these policies—termed as "low sustainability companies." In particular, they find that the boards of directors of these companies are more likely to be responsible for sustainability and top executive incentives are more likely to be a function of sustainability metrics. Finally, they provide evidence that HSCs significantly outperform their counterparts over the long term, both in terms of stock market and accounting performance.

Several firms that are considered HSCs are listed on the Dow Jones Sustainability Index and are earmarked by sustainability monitors and investors. They include Unilever (food), Akzo Nobel and DSM (chemicals), Philips (electronics), Puma and Nike (footwear), Ikea (furniture), and Google (internet). These companies are smartly leveraging a trend of sustainable behavior in society in general—that wave of eco-conscious regulators, health-conscious consumers, meaning- and balance-seeking employees, future-focused students, and increasingly influential faith-based and spiritual communities.[25]

This is not to say that these companies are sustainable in all aspects of their business. For example, Unilever, which is rated among peers as the leading sustainability company,[26] is still struggling with the transformation of various supply chains, especially in the detergent categories.[27] But taking a wider historical perspective, it is fair to say that most of these corporations are engaged in a process of *transforming themselves* into organizations that deliver sustainable value, beyond mere financial growth. Some take small steps, others larger steps, some stumble along the way, some leap forward. Yet we can view them as steps along the way for companies committed to sustainability, from which we can learn to create sustainable economic systems.[28]

25   Williams, F. (2015). *Green Giants: How Smart Companies Turn Sustainability into Billion-Dollar Businesses*. New York, NY: AMACOM.

26   For four years in a row until this report (in 2014), Unilever was regarded as the number one corporate sustainability leader, with expert respondents identifying the company as a "leader in integrating sustainability into its business strategy". See http://www.globescan.com/news-and-analysis/press-releases/press-releases-2014/310-global-ex.perts-rank-unilever-number-one-for-sustainability-leadership-in-new-survey.html.

27   Mirvis, P. (2011). Unilever's drive for sustainability and CSR—changing the game. In S.A. Mohrman, & A.B. Shani (Eds.), *Organizing for Sustainability Volume 1* (pp. 41-72). Bingley: Emerald Group Publishing Limited.

28   WBCSD (2011). *Collaboration, Innovation and Transformation: A Value Chain Approach*. Geneva: World Business Council for Sustainable Development;

9.2.1 Business sustainability is a process

As business sustainability is a process of change toward creating sustainable value, recent research has established that firms engaged in this process transition through a number of stages. At each stage there are discernable barriers and tipping points, where firms gradually take on a larger sustainable–relational worldview, with an enhanced sense of responsibility and stronger intrinsic motivation for sustainability.[29] Van Tulder has defined three stages in this process and four archetypes: inactive, reactive, active, and proactive.[30]

Companies leave the inactive phase typically in reaction to an external triggering event. Examples are the Brent Spar incident for Shell and the child labor "sweat shops" for Nike.[31] In the subsequent stage ("reactive"), firms will usually redefine and restructure their production process and supply chains, which will help them to move toward operational excellence. The sustainability expert Bob Doppelt speaks of this stage as the discovery of a new production paradigm: from "take–make–waste," which was common practice in the industrial era, to "borrow–use–return," which respects both the cyclicality of nature and social equity needed for a truly sustainable production process.[32]

The next stage ("active") typically involves a process of increased collaboration among stakeholders along the entire value chain, in order to meet

Nidumolu, R., Kramer, K., & Zeitz, J. (2012). Connecting heart to head: a framework for sustainable growth. *Stanford Social Innovation Review*, Winter. Retrieved from http://www.ssireview.org/articles/entry/connecting_heart_to_head; Kiron, D. (2013). The benefits of sustainability-driven innovation. *MIT Sloan Management Review*, 54(2), 69-73.

29   Mirvis, P. (2011). Unilever's drive for sustainability and CSR—changing the game. In S.A. Mohrman, & A.B. Shani (Eds.), *Organizing for Sustainability Volume 1* (pp. 41-72). Bingley: Emerald Group Publishing Limited; van Tulder, R., van Tilburg, R., Francken, M., & de Rosa, A. (2014). *Managing the Transitions to Sustainable Enterprise: Lessons from Frontrunner Companies*. London: Earthscan/Routledge.

30   van Tulder, R., van Tilburg, R., Francken, M., & de Rosa, A. (2014). *Managing the Transitions to Sustainable Enterprise: Lessons from Frontrunner Companies*. London: Earthscan/Routledge.

31   Visser, W. (2010). The age of responsibility: CSR 2.0 and the new DNA of business. *Journal of Business Systems, Governance and Ethics*, 5(3), 7-22.

32   Doppelt, B. (2009). *Leading Change Toward Sustainability: A Change-Management Guide for Business, Government and Civil Society* (2nd ed.). Sheffield, UK: Greenleaf Publishing.

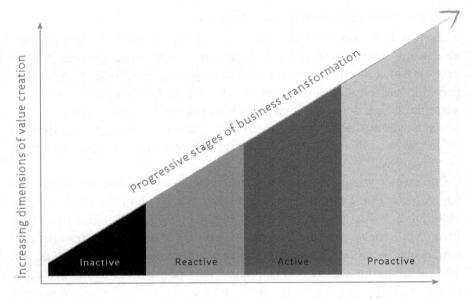

FIGURE 9.2   Stages in business sustainability

Source: Adapted from the phase model presented by Van Tulder et al. (2014).

their existing and future needs and to start to integrate the needs of society and nature. At this stage, companies begin to expand the amount of stakeholders with whom they are interdependent, such as financiers, NGOs, and end-consumers. Without involving the financial industry, many sustainability investments that are critically needed cannot be made. Likewise, without changing consumer behavior and taking into consideration projected population growth in the decades to come, sustainability efforts will fall short of reaching the necessary scale and impact. In any sustainability scenario, planetary boundaries will soon be crossed if we fail to adjust current consumption levels in developed nations downward. In other words, sustainability requires large-scale changes in consumer behavior.

At this stage companies will need to rethink the purpose of marketing. The exclusive focus on pushing demand from consumers to sell their products needs to move to a more balanced focus on serving their needs and creating a valuable relationship with a positive outcome for the outside world. Demand thinking tends to confuse needs with wants. While wants are limitless yet largely unnecessary for life satisfaction, needs tend to be more modest but are critical for well-being and happiness. In this regard, Unilever is actively experimenting with changing consumer behavior by launching a new shampoo that can be used without taking a shower, thus substantially lowering water usage, as described in the Unilever Sustainable Living Report.[33] In this case Unilever distinguished between the *need* for hygiene and sustainable water and the *want* for daily showers.

> The exclusive focus on pushing demand from consumers to sell their products needs to move to a more balanced focus on serving their needs and creating a valuable relationship with a positive outcome for the outside world.

In the final stage ("proactive"), companies will have fully embraced the sustainable–relational worldview through a process of co-creation with all stakeholders in the value chain with the aim of creating a collective positive impact on societal needs. Yet very few businesses have evolved to this stage. The reason for this is obvious: the existing paradigms keep them locked into a reactive mind-set, focused on their own needs and those of the stakeholders and top management. When things happen, they just do their best to hang on, and hopefully do better next time around. It takes all their energy to keep current business going, avoid losses, and make payroll. Yet some companies plan for the future—they embrace a proactive mind-set. They understand that context and markets shift, technology evolves, and unexpected waves of mayhem occur. These types of company tend to perform better than the reactive ones, because they have the mind-sets, resources, and foresight to weather the storm and create shared value. Their leadership has given thought to the murky future ahead and allocates resources to strategies for adapting to the future needs of society and how to serve them best. In this way, it becomes easy to see that the best companies are those that have high awareness of future changes, take control of their own future and the future of the society in which they participate. These insights are shown in Figure 9.2.

---

33   Unilever (2012). *Unilever Sustainable Living Plan: Progress Report 2012*. Retrieved from https://www.unilever.com/Images/uslp-progress-report-2012 -fi_tcm13-387367_tcm244-409862_en.pdf.

9.2.2 Creating shared value

> If you look at shopkeepers, although their main interest is profit, they
> know that the proper way to achieve it is not by force, but by smil-
> ing and a friendly attitude. ... In the long run, it is in the company's
> own interest to care about others—essentially, taking care of us all is
> in one's own interest.
>
> The Dalai Lama

There are many scholars studying the remarkable reorientation of business toward society and sustainability, with the Harvard strategy experts Michael Porter and Mark Kramer describing the process as "creating shared value" (CSV).[34] This concept emerged on the basis of transforming the value chain at Nestlé, in which the various players in a value chain collaboratively created more value than they would do when pursuing individual financial objectives.

The concept of CSV speaks to companies who want to go beyond the apparent dilemma between pursuing business results and creating social value, or between competition and compassion. They realize that by link- ing social and customer needs to their strategic organizational purpose, capabilities, and resources, they can enhance the long-term strategic com- petitive positioning and value creation of the firm. This linkage is made explicit through the vision of the company's leadership. Take, for example, the words of Unilever's CEO Paul Polman, when he explains why he has put sustainability at the top of his business agenda

> Most businesses operate and say how can I use society and the envi-
> ronment to be successful? We are saying the opposite—how can we
> contribute to society and the environment to be successful? So it
> starts with asking the right question to yourself, which will change the
> way you think.[35]

Similar comments have been made by Feike Sijbesma, the CEO of DSM, another firm regarded as a "high sustainability company": "As a business, we are aware that we cannot be successful in a society that fails. Therefore, it has become natural for us to take responsibility for more than our business,

---

34 Porter, M., & Kramer, M. (2006). Strategy & society: the link between competi- tive advantage and corporate social strategy. *Harvard Business Review*, 83(4), 66-67; Porter, M., & Kramer, M. (2011). The big idea: creating shared value. *Har- vard Business Review*, 1 (January–February), 64-77.

35 Forum for the Future (2011). Interview with Paul Polman. Retrieved from https://www.forumforthefuture.org/blog/6-ways-unilever-has-achieved- success-through-sustainability-and-how-your-business-can-too.

but also for society and nature."[36] As discussed in Chapter 5, similar concepts were employed by Ben Cohen and Jerry Greenfield, the founders of Ben & Jerry's, and by Bill George, the former CEO of Medtronic.

These comments reflect the concept of enlightened self-interest that was articulated at the first public dialogue, which so resonated with the Dalai Lama's views. In 1999, he talked of the main purpose of compassion: "Because it benefits us." He gave the example of how taking care of your customer with a genuine smile will help you gain more customers. He also clarified that compassion does not just mean charity or philanthropy: "If compassion were always to mean giving, then any company that acted compassionately would soon go bankrupt!" The Dalai Lama said that the best way to act compassionately is to empower people so that they don't become dependent on your compassion but be more constructive and compassionate themselves. This line of thinking corresponds to the idea of CSV. It is not about *sharing* the value already created by firms, which is the traditional redistribution approach underlying the notions of classical economics, corporate philanthropy, and CSR. Instead, CSV is about *expanding* the pool of economic and societal value.

The next stage is to explore how CSV can be put into practice. First and foremost, CSV is not only an analytical and strategic process.[37] It needs to be complemented by an organizational process that develops supportive worldviews, mind-sets, and capabilities among the leadership, as well as a rigorous measurement process to track and correlate societal, market, and business performance. In business currently, these processes tend to be run separately, but a genuine reorientation toward shared value creation requires these to be aligned.

9.2.3 New performance indicators

Why is measurement important in the transition toward CSV or business sustainability? Because there is a wide and vast array of environmental, social, and governance practices that don't meet—or don't easily fit into—the financial return expectations of today's shareholders and markets.

36   WBCSD (2012). *Measuring Impact Framework: A Guide for Business*. Geneva: World Business Council for Sustainable Development.

37   Porter, M., & Kramer, M. (2011). The big idea: creating shared value. *Harvard Business Review*, 1 (January–February), 64-77; Senge, P.M. (2008). *The Necessary Revolution: How Individuals and Organizations Are Working Together to Create a Sustainable World*. New York, NY: Broadway Books.

Shared value creation will not happen if companies continue to use conventional economic indicators as measures of their performance, be it at a national, corporate, or project level. Conventional indicators tend to merely measure financial results and not the value created with the different stakeholders. Moreover, most financial output indicators tend to misrepresent the real cost price for the value created or destroyed.[38]

However, equating value creation with financial output is like confusing the goal with the goalposts, or mixing up the means with the end. Organizations that do that, for example by focusing on maximizing short-term profits for their own needs, may end up destroying social value. The footwear company, Puma, recognized the danger of this approach and created an environmental profit and loss statement, accounting for impacts on water supplies from its production process.[39] Even though this was done on an experimental basis, it underscores the need, as was expressed in the discussion with the Dalai Lama, to "change the rules" of the current game. The basic logic that counts here is that you can only manage what you measure.

Logic implies that sustainability efforts can only succeed in the long term if they align business needs with customer and societal needs, while embedding this in the governance and reporting system. An encouraging development is that a number of states now allow corporations to be registered as so-called "benefit corporations," or B Corps. Some 1,800 B Corps in 50 countries have been so certified as of this writing.[40] B Corps are, generally, required to have a corporate purpose to create a materially positive impact on society and the environment. They provide for directors to have fiduciary duties to consider nonfinancial interests and they have an obligation to report on social and environmental performance assessed against an objective standard.[41]

New performance indicators will help companies to align the respective needs of customers, business, and society with the value creation capacity

38 Epstein, M.J. (2014). *Making Sustainability Work: Best Practices in Managing and Measuring Corporate Social, Environmental, and Economic Impacts.* Sheffield, UK: Greenleaf Publishing.

39 Nidumolu, R., Kramer, K., & Zeitz, J. (2012). Connecting heart to head: a framework for sustainable growth. *Stanford Social Innovation Review*, Winter. Retrieved from http://www.ssireview.org/articles/entry/connecting_heart_to_head.

40 For information about B Corps, see http://www.bcorporation.net.

41 Several states in the U.S.A., together with Belgium, Greece, and the U.K., have introduced special legal vehicles for enterprises, which aim to pursue a social purpose instead of making profit.

of human beings in organizations. In a sense, this equates with aligning firms' CSV intent and capacity to serve needs of stakeholders with actual societal needs. The exciting result will be that companies dramatically enhance their value creation capacity. Specifically, the value that is created expands in a number of stages: from financial value, to customer value, and to societal value. Since the shared value created spans three dimensions, in line with the "I," "Org," and "All" levels, it is more appropriate to describe it as "triple value."

The next question is how to develop a comprehensive measurement model that can capture this shared value creation. Since the current and new paradigms are expressed in different languages and models, we have looked for a theory of change model from the current transactional paradigm that can be easily adapted with concepts that are suitable to the new relational paradigm. One appropriate model turns out to be the Logical Framework Approach (LFA),[42] which speaks of three types of result—output, outcome, and impact—which are generated by inputs and activity. LFA is a well-known model that is applied in the world of assessing results from development assistance—when I worked on development projects in Tibet, I had to report to my donors on the basis of LFA.

Output is the direct result of a company's activities, such as number of products sold, financial profits, and employee satisfaction. Outcomes are the changes in human (consumer) behavior in markets, for example the amount of people who consume the product or not, or the awareness that a certain behavior, such as (to take the example of Unilever's Lifebuoy soap discussed in Chapter 5) washing your hands five times a day is good for personal hygiene. Impact is the effect on society as a result of all the outputs and outcomes combined, such as increased health of children due daily hand-washing with soap, or securing food security by reducing and recycling household food waste.[43] Ultimately, we want to generate not only a positive output, but also net positive outcomes and impact.

---

42 NORAD (1999). *The Logical Framework Approach (LFA): Handbook for Objectives-Oriented Planning* (4th ed.). Oslo, Norway: NORAD.

43 The World Bank defines food security as the state in which all people, at all times, have physical and economic access to sufficient, safe, and nutritious food that meets their dietary needs and food preferences for an active and healthy life. See World Bank (2014). *Food Price Watch*, 16. Retrieved from http://www-wds.worldbank.org/external/default/WDSContentServer/WDSP/IB/2014/05/30/000470435_20140530113813/Rendered/PDF/883900NEWS0Box000FP W0Feb020140final.pdf.

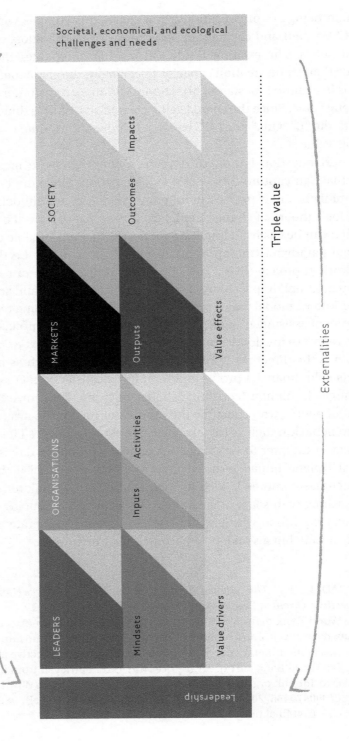

FIGURE 9.3   Triple value creation measurement framework

Among the myriad success drivers for a sustainability initiative, a key component is the design of the initiative itself.[44] In turn, one of the most important influences on the design of change initiatives is the worldview of the initiator or the leader.[45] Therefore, to represent the new reality in which leadership plays such an important role in shaping outcomes and impact, we have added leadership mind-sets as the foundational input in the LFA model. On the impact side, we have added societal, economic, and ecological needs (or current challenges defined in the Sustainable Development Goals)[46] as the ultimate goals. The beauty of this model is that it can actually measure and align all the steps in the value creation chain. In doing so, the model can help companies to move beyond the preoccupation with generating **financial capital** while ignoring its impact on **social capital** and **natural capital**. In other words, this framework can used for the creation of shared value and shared capital (Fig. 9.3).

Since this framework includes the dimensions of leadership ("I"), organizations ("Org"), and market and society ("All"), as well as its underlying chain of cause and effect expressed in value creation aimed at serving needs, it can be considered as one of the most holistic measurement models to date. It lends justice to both the outcomes of the discussions with the Dalai Lama and the actual business reality in today's complex environment. By placing leadership and its mind-sets at the foundation of the value creation process, it creates an exciting new perspective on the potential role of leadership in solving today's complex challenges. Now let us explore the domain of leadership for triple value creation in more depth.

44   Doppelt, B. (2009). *Leading Change Toward Sustainability: A Change-Management Guide for Business, Government and Civil Society* (2nd ed.). Sheffield, UK: Greenleaf Publishing; Kotter, J.P. (1996). *Leading Change*. Boston, MA: Harvard Business School Press.

45   Doppelt, B. (2009). *Leading Change Toward Sustainability: A Change-Management Guide for Business, Government and Civil Society* (2nd ed.). Sheffield, UK: Greenleaf Publishing; Sharma, S. (2000). Managerial interpretations and organizational context as predictors of corporate choice of environmental strategy. *Academy of Management Journal*, 43(4), 681-697.

46   United Nations (no date). Sustainable Development Knowledge Platform. Retrieved from https://sustainabledevelopment.un.org.

9.3 Leadership for triple value creation

> Leaders are not born or made, but cultivated and developed. The best
> way to develop leaders is through the training of values.
>
> The Dalai Lama[47]

As discussed in the dialogues with the Dalai Lama, the neuroscientific discovery about the potential of the human brain provides encouragement to
sustainability advocates. The classical economics ideal of the static *Homo
economicus* who is innately selfishness and individualistic has no foundation in biological reality and therefore can no longer serve as foundation for
economic theory.[48] Instead, there is ample evidence to show that the human
condition is full of positive potential: people have an intrinsic need and
drive to be supportive, creative, and generous, and they feel naturally linked
to their social and ecological context. As the Dalai Lama has said, when people are educated to develop qualities such as compassion, patience, power,
and wisdom, their minds will be more happy, balanced, and stable. This is
not a matter of ethics but of self-benefit; people who act in a balanced way
will be able to deal with a complex economic reality more effectively and
generate better results.

This promising insight serves as prelude to our discussion of leadership
development. Can we educate leaders so that they can create sustainable
organizations that generate shared value and not only "win" for themselves
but "win" with society? If so, what should be the principal elements of such
leadership education? What should be different from the existing concepts
behind leadership development? Can these ideas be expressed in a new
theory and practice for leadership? In order to define this leadership, in the
following discussion I will interchangeably use terms such as sustainability leadership, triple value leadership, future leadership and—later in the
chapter—societal leadership. First, we will need to reflect on what aspects
of leadership we need to look at.

As argued in the preceding pages, since the current problems and solutions can be traced back to the concepts, paradigms, beliefs, and values that
we hold in our mind, or in other words **our mind-sets**, I believe that we need
to look specifically at what the **mind-sets of future leaders** are: where do

47 Singer, T., & Ricard, M. (2015). *Caring Economics: Conversations on Altruism
and Compassion Between Scientists, Economists, and the Dalai Lama*. New York,
NY: Picador.

48 Gintis, H. (2000). Beyond homo economicus: evidence from experimental economics. *Ecological Economics*, 35, 311-322.

they arise from and how can they be grown and expanded? The term mind-set, as I use it, refers to interior patterns of mind or frames of reference. Scholars use several terms interchangeably when referring to mind-sets, including worldviews, meaning-making systems, action logics, and mental models.[49] The mind-set of leaders determines how they see the world, how they reason, and how they make meaning and behave in response to their experiences in the world.[50]

The "Compassion or Competition" dialogues have revealed a number of recurrent themes that hold relevance to the mind-sets of the leadership of the future. In essence, the themes represent different ways of thinking from the current economic paradigm. As indicated in the previous chapters, these themes can be distilled into the following six categories:

1. **Holistic thinking.** Reality can be seen as interconnected and interdependent at various levels of complexity, including the personal ("I") dimension, the organizational ("Org") dimension, and the societal, economic, and ecological systems dimension ("All"). This requires a holistic approach to satisfying needs.

2. **Compassion.** People are naturally wired for compassion, and people in business are no exception. In fact, business cannot exist without giving services to, and receiving services from, society. As the needs of society change and grow, the role of business is to serve those needs in collaboration with many stakeholders.

3. **Responsibility.** Leaders have the capacity to develop a proactive and responsible attitude, regardless of the circumstances. In other words, they can "step into their power." This attitude leads to self-confidence and courage, and allows for resilience and a sense of contentment in the face of difficult outer challenges.

4. **Creativity.** With a fast and fundamentally changing context, it has become important to actively seek solutions, which can only be co-created with others. This requires leaders to develop a collective purpose with long-term goals, while allowing for experimentation and fast prototyping.

49 Schein, S. (2015). *A New Psychology for Sustainability Leadership: The Hidden Power of Ecological Worldviews.* Sheffield, UK: Greenleaf Publishing.
50 Dweck, C.S. (2006). *Mindset: How You Can Fulfill Your Potential.* London: Robinson.

5. **Collectiveness.** Since everything is increasingly interconnected, solutions for the future are beyond individual effort. This may include changing the current rules of the game, new governance structures, new measurement indicators that go beyond short-term and financial measurement, establishing collaborative structures, and improving resources and ground structures such as education.

6. **Education of the mind/heart.** It is important and possible to continuously train the mind and heart, for example through meditation and cultivating positive attitudes. The learning process involves a full engagement of our physical, emotional, mental, and spiritual qualities, rather than the mere training of our cognition. In this way, levels of leadership mind-sets and qualities can be developed and grown.

9.3.1 Toward a framework for triple value leadership

In order to translate these themes into the leadership mind-sets that are needed in the future, I interviewed 25 leaders in business who were considered eminent in the fields of sustainability and social innovation, including CEOs, social entrepreneurs, chief CSR/sustainability officers, and thought leaders. In addition, my team and I studied the leadership styles of organizations with sustainability in their competitive positioning, such as Unilever, DSM, Nike, Patagonia, Triodos Bank, and Rabobank.

I also reviewed the academic literature on leadership over the last few decades and focused specifically on the growing field of leadership for sustainability.[51] This field goes by many different names, depending on the

51 Senge, P.M. (2008). *The Necessary Revolution: How Individuals and Organizations Are Working Together to Create a Sustainable World.* New York, NY: Broadway Books; Lueneburger, C., & Goleman, D. (2010). The change leadership sustainability demands. *MIT Sloan Management Review,* 51(4), 49-55; Gitsham, M. (2009). *Developing the Global Leader of Tomorrow.* Berkhamsted, UK: Ashridge Business School; Johansen, B. (2012). *Leaders Make the Future: Ten Leadership Skills for an Uncertain World.* San Francisco, CA: Berrett-Koehler Publishers; Marshall, J., Coleman, G., & Reason, P. (2011). *Leadership for Sustainability: An Action Research Approach.* Sheffield, UK: Greenleaf Publishing; Nidumolu, R., Kramer, K., & Zeitz, J. (2012). Connecting heart to head: a framework for sustainable growth. *Stanford Social Innovation Review,* Winter. Retrieved from http://www.ssireview.org/articles/entry/connecting_heart_to_head; Doppelt, B. (2012). *The Power of Sustainable Thinking: How to Create a Positive Future for the Climate, the Planet, Your Organization and Your*

perspective it addresses, including corporate social responsibility (CSR) leadership, environmental leadership, transformational leadership, and ethical leadership. The most relevant dimensions of this literature for my study are those that identify the mind-sets,[52] competences,[53] and the behaviors[54] that sustainability leaders need.

Most of this research is exploratory. Nonetheless, some studies strongly support the need for leaders that have a sophisticated worldview and have begun to document what such a perspective looks like in practice.[55] Metcalf and Benn define this as follows: "Leadership for sustainability requires leaders of extraordinary abilities. These are likely to be leaders who can read and

Life. London: Earthscan; Mackay, J., & Sisodia, R. (2013). *Conscious Capitalism: Liberating the Heroic Spirit of Business*. Boston, MA: Harvard Business Review Press; Schein, S. (2015). *A New Psychology for Sustainability Leadership: The Hidden Power of Ecological Worldviews*. Sheffield, UK: Greenleaf Publishing; Metcalf, L., & Benn, S. (2013). Leaders for sustainability: an evolution of leadership ability. *Journal of Business Ethics*, 112(3), 369-384.

52 Boiral, O., Cayer, M., & Baron, C.M. (2009). The action logics of environmental leadership: a developmental perspective. *Journal of Business Ethics*, 85, 479-499; Shrivastava, P. (1994). Ecocentric leadership in the 21st century. *The Leadership Quarterly*, 5(3), 223-226; Brown, B.C. (2011). *Conscious Leadership for Sustainability: How Leaders with a Late-Stage Action Logic Design and Engage in Sustainability Initiatives* (Unpublished PhD dissertation). Fielding Graduate University, Santa Barbara, CA.

53 Hind, P., Wilson, A., & Lenssen, G. (2009). Developing leaders for sustainable business. *Corporate Governance*, 9(1), 7-20; Kakabadse, N.K., Kakabadse, A.P., & Lee-Davies, L. (2009). CSR leadership road-map. *Corporate Governance*, 9(1), 50-57.

54 Doppelt, B. (2009). *Leading Change Toward Sustainability: A Change-Management Guide for Business, Government and Civil Society* (2nd ed.). Sheffield, UK: Greenleaf Publishing; Quinn, L., & Dalton, M. (2009). Leading for sustainability: implementing the tasks of leadership. *Corporate Governance*, 9(1), 21-38.

55 Boiral, O., Cayer, M., & Baron, C.M. (2009). The action logics of environmental leadership: a developmental perspective. *Journal of Business Ethics*, 85, 479-499; Doppelt, B. (2009). *Leading Change Toward Sustainability: A Change-Management Guide for Business, Government and Civil Society* (2nd ed.). Sheffield, UK: Greenleaf Publishing; Hames, R.D. (2007). *The Five Literacies of Global Leadership: What Authentic Leaders Know and You Need to Find Out*. Hoboken, NJ: Jossey-Bass; Hardman, G. (2009). *Regenerative Leadership: An Integral Theory for Transforming People and Organizations for Sustainability in Business, Education, and Community* (Unpublished PhD dissertation). Florida Atlantic University, Boca Raton, FL; Brown, B.C. (2011). *Conscious Leadership for Sustainability: How Leaders with a Late-Stage Action Logic Design and Engage in Sustainability Initiatives* (Unpublished PhD dissertation). Fielding Graduate

Mind-set	Theme	Concepts used in literature and interviews	Question
Context awareness	Holistic thinking	Extending perspective; dealing with complexity; recognizing interconnectedness; interdependence; continuity; systems thinking; ambiguity; dilemmas; the middle way; recognizing mega-trends and key changes in value chains and larger society	Where
Connectedness	Compassion	Recognizing and serving needs of all stakeholders; social and emotional skills; empathy; building valuable relationships; navigating social tensions and paradoxes; collaboration; fairness, care, altruism; courage	Who
Centeredness	Responsibility	Self-knowledge; calm and stable, grounded; resilience, self-confidence, responsibility, purpose; meaning; structure; resilience; autonomy; self-determination; adaptation	Why
Collective vision	Creativity	Long-term horizon; creating a vision; alignment; engagement inspiration; goal-setting; creation of shared value; solving societal and business challenges simultaneously; impact; optimism	What
Creative competence	Collective effort	Rules of the game; output, outcome and impact; open source; competitive activities, resources, competences; value creation; embedding sustainability in business creation process and value chains; governance; measurement and performance	How
Consciousness	Education	Mind-sets; worldviews; mental model; beliefs; moral compass; values; attitudes; paradigms; presence; open-mindedness; self-awareness; continuous learning; clarity of mind; mind-training; suspending judgment	Within

TABLE 9.1   The 6C mind-sets

predict through complexity, can think through complex problems, engage groups in dynamic adaptive organizational change and can manage emotion appropriately."[56]

Bringing these various sources together, I have found six specific mind-sets that could make up the leadership typology of the future:

1. Context awareness

2. Connectedness

3. Centeredness

4. Collective vision

5. Creative competence

6. Consciousness

I call these the "6C" mind-sets (Table 9.1).

The most important mind-set is the capacity to recognize and effectively relate to rapidly changing context in the world. This includes an ability to understand and handle increasing levels of complexity and systemic interdependence between actors in their value chains. I call this "**context awareness.**" The Dalai Lama describes this as holistic thinking, the ability to extend one's perspective and broaden one's view, as well as the awareness that change is needed because reality is changing. In addition, this type of leadership requires a long-term view and a sense of continuity and interdependence while demonstrating open-mindedness.

Another important leadership mind-set is the willingness to connect to a wide range of others, especially the stakeholders of one's own organization—I call this "**connectedness.**" Leaders need to be equipped with advanced social and emotional skills to relate to and connect with multiple stakeholders, maintaining excellent relationships while navigating social tensions and paradoxes. Most important here is to recognize the different needs, interests, and views of stakeholders, both close to the organization and further away. In the view of the Dalai Lama, this is the practice of compassion, which starts with a sense of empathy or emotional resonance, but then moves into the logical analysis of the cause and conditions that lead to

University, Santa Barbara, CA; Metcalf, L., & Benn, S. (2013). Leaders for sustainability: an evolution of leadership ability. *Journal of Business Ethics*, 112(3), 369-384.

56   Metcalf, L., & Benn, S. (2013). Leaders for sustainability: an evolution of leadership ability. *Journal of Business Ethics*, 112(3), 369-384.

the problems at hand. And then finally to courage: the resolution to change, if possible.

This, in a sense, builds to the next leadership quality, which I call "**centeredness**," the ability to stay calm and focused in the face of external challenges with a high degree of self-knowledge and willingness to continually learn. This really is a leadership attitude that generates resilience, self-confidence, and responsibility, and is linked to a sense of purpose or direction. It also related to the recognition that everyone (leader or no leader) has the capacity to carve out some space for making a contribution, however small, and develop a sense of greater purpose and engagement for the larger world. Everyone and every little action counts, said the Dalai Lama, and one should not be discouraged if the results are not immediately visible and tangible.

The next crucial quality of leaders is what I call "**collective vision**," the ability to develop a vision for the future, one that inspires, aligns, and engages others, and stimulates co-creation with all stakeholders in order to unlock new innovative sustainable strategies. I am not speaking of some sort of faraway utopian ideal, but of a vision that is realistic and achievable ("realistic action," said the Dalai Lama) so that sustainable ideas and strategies can be turned into shared value creation. In other words, future leaders need to possess a strong collaborative and visionary mind-set, guided by societal needs, without losing touch with where we stand today.

In addition, in today's rapidly changing and interconnected world, leaders need to be able to create different values for different needs, for example the financial needs of shareholders, the employment and health needs of employees, the service needs of customers, and the well-being needs of society. That is, they need to become creative and at the same time focus on the unique competences, resources, and capabilities of the firm to generate competitive activities that create (rather than destroy) value. I call this "**creative competence**." This includes a commitment to embed sustainability into organizational strategy and structures, change the rules of the game in the value chain, and establish internal and external performance and shared value creation measures in a comprehensive and guiding framework. In this regard, the Dalai Lama spoke of leadership beyond individual effort, and focusing on solutions that generate collective and lasting benefit.

Finally, the Dalai Lama repeatedly called for educating our mind and heart, as a practice for everyone, especially leaders, a call that resonated with many others in the field of sustainability and leadership. He emphasized this point by referring to both Buddhist practice and recent insights from neuroscience that demonstrate that we can constantly develop our brain (a phenomenon called neuroplasticity) and expand our consciousness, our mind, and frame of reference. Some call this the development of a mental state of fully developed awareness; I call it "**consciousness**."[57]

The 6C mind-sets correspond with the six themes derived from the dialogue with the Dalai Lama. Since my aim is to develop practical (i.e., easy to remember) management concepts, I have also incorporated the familiar questions, "where," "why," "who," "what," "how," and "within" (Table 9.1), the logic of which I will explain in Section 9.4.

9.3.2 Leadership mind-set development and purpose

Now that we have established which mind-sets leaders need to develop, the next question that arises is: how can these mind-sets be developed and grown? How can leaders enhance their mind-set capacity to create value for themselves and their environment—in other words, how can they grow as leaders? These questions are researched in several academic disciplines, especially **leadership psychology**, intersecting with various other fields, including **neuroscience** and branches of (social) psychology, organizational dynamics, and sustainability.

Researchers in the field of **developmental theory** have studied how people manifest their potential as they grow up into maturity.[58] Developmental theories provide a coherent framework to understand how people interpret events and their experiences, and can also predict how people are likely to act in many situations. They have also identified a number of stages through which people progress when they grow and develop their mind-sets. Roughly, each of the stages of development involves the reorganization of meaning-making, perspective, self-identity, and the overall way of knowing. Similar to scaling a mountain, the higher one climbs, the further one can see.

The psychologist Carol Dweck has observed that there are broadly two mind-sets that present a whole world of difference: fixed mind-sets versus

---

57   Senge, P., Sharmer, O., Jaworski, J., & Flowers, B.S. (2008). *Presence: Human Purpose and the Field of the Future.* Boston, MA: Crown Business.

58   Beck, D.E., & Cowan, C.C. (1996). *Spiral Dynamics: Mastering Values, Leadership, and Change.* Cambridge, MA: Blackwell; Dweck, C.S. (2006). *Mindset: How You Can Fulfill Your Potential.* London: Robinson; Kegan, R. (1982). *The Evolving Self: Problem and Process in Human Development.* Cambridge, MA: Harvard University Press; Fisher, D., Rooke, D., & Torbert, W. (2003). *Personal and Organizational Transformations Through Action Inquiry.* London: The Cromwell Press.

growth mind-sets.[59] Fixed mind-sets are, as the term indicates, static, and occur when one tries to fit the world into one's set of beliefs and ideas. This provides a sense of short-term security, but in the longer term a fixed mind-sets causes a lot of problems, as the reality tends to outgrow the mind-set. Growth mind-sets, on the other hand, provide flexibility and the opportunity to learn. Circumstances in life are regarded as an invitation to grow and develop. People with growth mind-sets tend to be more successful and content in their lives and work. Interestingly, there is no direct relationship between growth mind-sets and higher levels of IQ.[60]

The question here is: to what extent are people able to grow and expand their own perspectives and frames of reference? What is the potential of leaders **to shift and transform their mind-sets**, so that they will be better equipped to deal with the challenges of today? Abraham Maslow discovered this process of development and called it a process of recognizing one's "higher-order needs": once people have satisfied "lower-order needs" such as security and self-esteem, they can start serving other needs, such as the need to take responsibility, to achieve results, to develop mutually satisfying relationships, to develop oneself and others, and to continuously improve.[61]

The notion of psychological empowerment has been specifically studies in the school of **motivational psychology**, which explores how motivation manifests in an individual's orientation to his or her life or work.[62] The neuroscientist Tania Singer has found neurological correlates for different motivational drivers, as each of them is activating different parts of the brain. Importantly, she has found that people have the capacity to voluntary activate intrinsic motivational drivers, including the affiliative-caring motive related to serving others.[63]

This provides a biological basis for people to enhance their sense of purpose and empowerment; the psychologists Edward Deci and Richard Ryan,

59   Dweck, C.S. (2006). *Mindset: How You Can Fulfill Your Potential.* London: Robinson.

60   Dweck, C.S. (2006). *Mindset: How You Can Fulfill Your Potential.* London: Robinson.

61   Maslow, A.H. (1943). A theory of human motivation. *Psychological Review*, 50(4), 370-396.

62   McClelland, D.C. (1978). Managing motivation to expand human freedom. *American Psychologist*, 33(3), 201-210.

63   Singer, T. (2008). Understanding others: brain mechanisms of theory of mind and empathy. In P.W. Glimcher, C.F. Camerer, E. Fehr, & R.A. Poldrack (Eds.), *Neuroeconomics: Decision Making and the Brain* (pp. 233-250). Amsterdam, Netherlands: Elsevier.

in their well-researched self-determination theory, define empowerment as a process of pursuing goals based on the fulfillment of the need for autonomy, relatedness, and competence in relation to the context.[64]

When autonomy, relatedness, and competence are present in individuals, they can undertake a specific goal or challenge according to the context. When these three attributes are present, the individual will experience a sense of energy and flow, which corresponds closely to the theory of "flow" put forward by the psychologist Mihaly Csikszentmihalyi when describing the experience of people when the demands of the activity are in balance with one's capacity.[65] When people's internal capacity is equipped for dealing with the external challenge, the latter then becomes what is called an "optimum challenge." The resulting experience is one of "flow," which in this context I define as "purpose." Importantly, the sense of purpose gives rise to a sense of power—or better, a sense of empowerment.

The study of purpose and meaning-making in leadership has evolved into a field known as the **constructive–developmental theory**.[66] One exponent of this field, Susanne Cook-Greuter, describes two main aspects of mindset growth and expansion: horizontal (or lateral) development and vertical development.[67] Horizontal development refers to the expansion in capacities through an increase in knowledge, skills, and abilities associated with a current mind-set. Vertical development is associated with capacity shifts from an

64 Deci, E.L., & Ryan, R. (2000). The self-determination theory, the facilitation of intrinsic motivation, social development, and well-being. *American Psychologist*, 55(1), 68-78.

65 Csikszentmihalyi, M. (1990.) *Flow: The Psychology of Optimal Experience*. New York, NY: Harper & Row.

66 Cook-Greuter, S.R. (1999). Postautonomous ego development: a study of its nature and measurement. *Dissertation Abstracts International*, 60 06B (UMI No. 993312); Cook-Greuter, S.R. (2004). Making the case for a developmental perspective. *Industrial and Commercial Training*, 36(7), 277-281; Kegan, R. (1982). *The Evolving Self: Problem and Process in Human Development*. Cambridge, MA: Harvard University Press; Kegan, R. (1994). *In Over Our Heads: The Mental Demands of Modern Life*. Cambridge, MA: Harvard University Press; Loevinger, J. (1976). *Ego Development: Conceptions and Theories*. San Francisco, CA: Jossey-Bass; Fisher, D., Rooke, D., & Torbert, W. (2003). *Personal and Organizational Transformations Through Action Inquiry*. London: The Cromwell Press; Torbert, W.R., Cook-Greuter, S.R., Fisher, D., Foldy, E., Gauthier, A., Keeley, J., ... Tran, M. (2004). *Action Inquiry: The Secret of Timely and Transforming Leadership*. San Francisco, CA: Berrett-Koehler Publishers.

67 Cook-Greuter, S.R. (2004). Making the case for a developmental perspective. *Industrial and Commercial Training*, 36(7), 277-281.

individual's current way of meaning-making to a broader, more complex, and deeper mind-set. Vertical development describes **the increase of individual awareness**, or **consciousness**, the expansion of **what an individual can pay attention to**, and, therefore, what he or she can influence. Vertical development, therefore, leads to a greater degree of capacity and responsibility, and a greater sense of empowerment of one's purpose. An individual purpose can grow into an organizational purpose, which in turn can grow into a societal purpose, which I call a "shared purpose." When people develop vertically (that is their consciousness), they can ultimately be of more value to others.

The concept of shared purpose is recognized as an important leadership attribute in the history of leadership. This is particularly the case in the field of **transformational leadership**, which is a concept that is contrasted with transactional leadership.[68] While the latter defines leadership as skills and knowledge aimed at making people work effectively within the current status quo, transformational leadership helps others advance to a stronger intrinsic motivation and higher purpose. The leadership scholar Peter Block refers to this as the ability of leaders to "enact a purpose of greatness," which creates clarity and direction for more than oneself, for others who recognize themselves in the purpose.[69] Jim Kouzes and Barry Posner define leadership in similar terms: "Credible leaders choose to give power away in service of others and for a purpose larger than themselves. As a result, they become power generators from which others draw energy."[70]

In a recent empirical study, Barrett Brown has brought all these lines of thinking together, exploring the connection between purpose, leadership, and sustainability.[71] He conducted an empirical study of leaders from business, government, and civil society engaged in sustainability initiatives. The results provide a granular view of how such individuals may think and behave with respect to complex change initiatives. The leaders in this study appear to exhibit the three general qualities, which he called being,

68 Burns, J.M. (1978). *Leadership*. New York, NY: Harper & Row; Bass, B.M. (1998). *Transformational Leadership: Industrial, Military, and Educational Impact*. Mahwah, NJ: Lawrence Erlbaum Associates.

69 Block, P. (1987). *The Empowered Manager*. San Francisco, CA: Jossey-Bass.

70 Kouzes, J.M., & Posner, B.A. (1995). *The Leadership Challenge*. San Francisco, CA: Jossey-Bass. This line of thinking is also expressed in Fox, E.A. (2013). *Winning From Within: Breakthrough Method for Leading, Living, and Lasting Change*. New York, NY: HarperCollins.

71 Brown, B.C. (2011). *Conscious Leadership for Sustainability: How Leaders with a Late-Stage Action Logic Design and Engage in Sustainability Initiatives* (Unpublished PhD dissertation). Fielding Graduate University, Santa Barbara, CA.

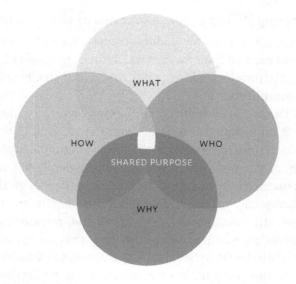

FIGURE 9.4   Shared purpose

reflecting, and engaging. "Being" refers to fundamental or essential qualities of these individuals, i.e., it has to do with characteristics of **who they
are**. "Reflecting" concerns **how they think** about and gain insight into the
design of change. "Engaging" addresses the **actions they take** to develop
and manage the design of change.

It is not difficult to translate these qualities into the mind-sets of the 6C
model: "being" relates to consciousness and centeredness, "reflecting" to
context awareness, and "engaging" to connectedness, collective vision,
and creative competence. The 6C model can also be recognized in the self-
determination theory: autonomy can also be described as centeredness
through being self-confident, courageous, and inwardly calm and stable—
they have a sense of purpose, they know "why" they live; relatedness is the
same as connectedness, which is facilitated by an altruistic mind-set, for
relatedness allows them to know with "whom" they are related; and competence represents the ability to achieve the desired results—the "how."

Obviously, all these qualities are connected. As the "why" deepens, the
"what" and the "who"/"how" can expand in scope and depth as well. Most
fundamental is the inner mental state of the leader, centered in the "why"
or the purpose of the leader. Seen within the leadership perspective where
leaders develop to pursue a purpose bigger than their own benefit, their
purpose can become a higher or "shared purpose." The way that purpose
arises is depicted in Figure 9.4.

9.3.3 Connecting shared purpose to levels of complexity

The question is not merely how leaders can discover and develop their sense of purpose and expand their awareness. Rather, the point of our inquiry is that they need to develop their capacity for creating triple value at increasing levels of complexity.

The Dalai Lama has identified the various levels of complexity in the dialogues. It is important to note that he never referred to them as merely abstract levels of thinking. For him, the levels of complexity are features of the principles of interconnectedness and interdependence, which are central to the business reality of today. The key lies in the recognition that, since we are fundamentally interconnected and interdependent with everyone on this planet, the development of an altruistic, responsible mind-set is actually in accordance with one's own interest. This is what we described in the dialogues with the Dalai Lama as "enlightened self-interest." On the one hand, the principle of interconnectedness requires leaders to think holistically and to consider their actions at multiple levels of complexity. On the other, these levels represent three stages in the motivation of the leader—to which level of stakeholder do leaders extend their motivation of care? From this it follows that we can translate the levels of complexity into three levels of consciousness:

1. Responsibility and care for oneself ("I")

2. Responsibility and care for others ("Org")

3. Responsibility and care for all ("All")

In this way, the three levels can be regarded as progressive stages in leadership development.

First, leaders need to be able to distinguish the three levels of complexity and the stakeholders involved. Second, they need to have the capacity and competence to serve the needs of these different stakeholders. It starts with taking responsibility and care for themselves, such as developing their intellectual, physical, spiritual, and emotional intelligence. This will enable them to deal with their own minds, their emotional well-being, and their health.

The second level is care for others. In the case of a leader, this means care for one's direct reports, one's team and support staff. Beyond that, leaders need to extend their responsibility and care throughout the organization that they serve.

While most leaders are aware of these first two levels, the third—care and responsibility for all—is less commonly practiced even though it is becoming more relevant in today's complex world. It is the level of care for society beyond the organization, and can be divided into three broad groups: the value chain (customers, suppliers, investors, and regulators—also known as primary stakeholders—and including the sector, industry, and competitors); society (secondary stakeholders, including communities and civil society); and ecosystems (nature and the environment, biodiversity, next generations).

The three levels of consciousness correlate closely with six perceived levels of leadership:

"I"	Personal leadership	1. Personal leadership
"Org"	Organizational leadership	2. Team leadership
		3. Organizational leadership
"All"	Societal leadership	4. Market leadership
		5. Societal leadership
		6. Ecological leadership

Societal leadership can therefore be defined as the leadership that delivers sustainable value at the highest level of complexity, including market value, societal value, and ecological value. In other words, in order to solve complex sustainability challenges, we need to cultivate societal leadership in business.

9.3.4 The practice of developing societal leadership

> All beings want to live a happy life, but only human beings have the ability to analyze its causes. Although the mind is fundamentally pure, when we develop anger, attachment or even compassion, it seems that the mind is suffused with that emotion. Anger is a disturbing emotion based on wrong view, the misconception of things having intrinsic existence, itself based on the way they appear. Constructive emotions arise on the basis of a correct view of reality.
>
> The Dalai Lama[72]

Now that we have established the content of the mind-sets of societal leadership and the degrees of complexity that they need to grapple with, we

[72] Dalai Lama, & Norman, A. (2011). *Beyond Religion: Ethics for a Whole World.* New York, NY: Houghton Mifflin Harcourt Publishing.

need to understand what this means for the development and practice of societal leadership. How can we develop leaders with this exceptional repertoire of societal leadership capacity? What techniques and practices are available to leaders who want to develop societal leadership?

According to the Dalai Lama, human beings have the capacity to change their minds and attitudes toward life. While the process of changing your mind is often associated with the practice of meditation, the Dalai Lama made it clear that the main point of meditation is to generate a positive, responsible, and constructive open mind, with purpose, compassion, and courage. On this basis, action to create positive impact can be taken. In his mind, the word "meditation" means the training of one's mind by cultivating positive values and training pro-social qualities, which lead to a responsible and proactive attitude in life.[73]

In order to fully understand what the Dalai Lama means with this, it is worth exploring the source of his thinking, which is Buddhist philosophy and principles of mind-training. However, while in the Buddhist context mind-training has a specific spiritual goal, the Dalai Lama has stressed the importance and the possibility of cultivating the mind in a secular context, such as in education and business, while using the same principles of mind-transformation. What are these principles? As I described in Chapter 5, the best reference for this is the ideal of the Bodhisattva, which can be regarded as the Buddhist notion of leadership, for a Bodhisattva is someone who, primarily through the development of wisdom and compassion, decides to take on the path of mental development while encouraging others to do likewise.[74] Throughout Asian history, the Bodhisattva ideal has inspired many leaders such as statesmen, merchants, scholars, and artists.[75]

In the Buddhist view, wisdom is the correct understanding of how the outer and inner phenomenological world exists and operates, namely as an interconnected system.[76] In view of the interconnected nature of reality,

73 Dalai Lama (2002). *The Meaning of Life from a Buddhist Perspective*. Boston, MA: Wisdom Publications.

74 Shantideva (1997). *A Guide to the Bodhisattva Way of Life* (V.A. Wallace & B.A. Wallace, Trans.). Ithaca, NY: Snow Lion.

75 Loizzo, J. (2006). Renewing the Nalanda legacy: science, religion and objectivity in Buddhism and the West. *Religion East & West*, 6, 101-121; Loizzo J. (2012). *Sustainable Happiness: The Mind Science of Well-being, Inspiration and Compassion*. New York, NY: Routledge; Thurman, R. (1997). *Inner Revolution: Life, Liberty and the Pursuit of Real Happiness*. New York, NY: Riverhead Books.

76 Dalai Lama (2002). *The Meaning of Life from a Buddhist Perspective*. Boston, MA: Wisdom Publications.

the Bodhisattva does not make a distinction between self and others. While their principal focus is the development of their own mind, Bodhisattvas are equally concerned with the world around them. Given that others and we are interconnected and interdependent, and that our happiness in the ultimate sense relies on the happiness of others, compassion is considered a natural state. Bodhisattvas, therefore, work on their mind to develop wisdom while practicing compassion in finding ways to diminish the suffering of others. Their qualities of wisdom and compassion will need to be supported by a powerful, determined, and focused mind, brought about by meditation. Hence, we can say that the Bodhisattva has essentially three qualities: wisdom, compassion, and power.

In view of the fact that sustainability is driven by the principles of increasing interconnectedness and interdependence, and the Bodhisattva model is rooted in the same principles of interconnectedness, it can serve as an appropriate framework for sustainability leadership. Looking at the many leadership theories that have evolved over time, the Bodhisattva model and its underlying philosophy of interdependence may in fact be best suited for the development of leadership capable of dealing with the unprecedented complex challenges of today.

This chapter is not the place to discuss the Bodhisattva model in depth; suffice to say that it has encouraged me to seek corresponding insights in Western science.[77] The **self-determination theory** had already discovered that people's minds have an innate tendency toward their potential development and integrated functioning.[78] In order to actualize their potential, they need nourishment from the social environment on the basis of balancing the why, who, how, and what. If this happens, they will experience well-being and growth. When their intrinsic needs are met, people feel empowered to achieve goals appropriate to and necessary for the context in which they live. In other words, we are equipped with a natural regulatory process aimed at attaining psychological integration and well-being suited to the challenges of the context.

77   For a discussion on the Bodhisattva model in relation to modern leadership, see Tideman, S.G. (2016). Gross national happiness: lessons for sustainability leadership. *South Asia Journal for Global Business Research*, 5(2), 190-213.

78   Deci, E.L., & Ryan, R. (2000). The self-determination theory, the facilitation of intrinsic motivation, social development, and well-being. *American Psychologist*, 55(1), 68-78.

Other schools of research have also studied this regulatory process, coming together in the rapidly evolving field known as **contemplative science**,[79] which investigates the mind by combining objective empirical research (second/third-person perspective) with subjective experience (first-person perspective). The field includes the schools of psychology and neuroscience, as well as clinical approaches in medicine, psychiatry, and education. By including the first-person perspective, the research outcomes can be regarded as presenting a fuller picture of the human mind.

One of the approaches that has been extensively studied in the last decade is **mindfulness**, starting with MIT medical researcher Jon Kabat-Zinn who developed a program called mindfulness-based stress reduction (MBSR).[80] The neuroscientist Richard Davidson shows that this process can be observed in measurable change patterns in the brain[81] and, since then, multiple research studies have shown that this process of mind-regulation results in positive effects on one's mental and physical health and well-being.[82]

The result is that mindfulness has become a recognized method for enhancing people's health and well-being. More recently, these mindfulness-based methods have found application in organization and leadership. When these practices are complemented with other educational methods, such as dialogue and mind/body practice, they can become more than tools for people's sense of well-being: they help people to expand their awareness of oneself and one's environment.[83] In short, mindfulness turns out to be an instrument for personal regulation and people development in the workplace.

How does that work? Neuroscientists have observed that people can learn to shift out of preoccupation within oneself and one's own immediate concern. The technical language for this is the self-referencing network

79   Varela, F., Thompson, E.T., & Rosch, E. (1992). *The Embodied Mind: Cognitive Science and Human Experience.* Boston, MA: MIT Press; Wallace, B.A. (2006). *Contemplative Science: Where Buddhism and Neuroscience Converge.* New York, NY: Columbia University Press.

80   Kabat-Zinn, J. (1990). *Full Catastrophe Living: Using the Wisdom of Your Body and Mind to Face Stress, Pain, and Illness.* New York, NY: Bantam Books.

81   Davidson, R.J., & Begley, S. (2012). *The Emotional Life of Your Brain: How its Unique Patterns Affect the Way You Think, Feel, and Live—and How You Can Change Them.* New York, NY: Hudson Street Press.

82   Davidson, R.J., & Begley, S. (2012). *The Emotional Life of Your Brain: How its Unique Patterns Affect the Way You Think, Feel, and Live—and How You Can Change Them.* New York, NY: Hudson Street Press.

83   Siegel, D. (2009). *Mindsight: The New Science of Personal Transformation.* New York, NY: Random House.

that more or less automatically controls the minds of untrained people.[84] Through mindfulness, the default network can be short cut into states of mind that are more open to the concerns of others and the larger environment. Scientists speak of this openness—which correlates to an increased coherence in the neural network of the brain—as the open loop of our brain which results in a state of presence, or heightened awareness.[85] Brain coherence has also been observed in groups, for example jazz musicians in concert, which has been described as a state of "flow."[86] In this situation, individual brains are interlocked into an interpersonal experience of energy flow, creating a field of resonance among the group members.

The neuro-psychiatrist Dan Siegel describes this process as "integration," which is considered the prime function of a healthy mind.[87] Integration, as Siegel defines it, is not merely an internally oriented process within the brain, but both embodied and interpersonal, which means that it is shared through relationships with others and with the world. The process of integration in fact entails a process of recognizing one's natural interconnectedness with the world. These findings lend some credence to Mahatma Gandhi's saying, "Be the change you want to see in the world."[88]

Development psychology expressed this development as moving through the stages of dependent (child), independent (adolescent), autonomous (young adult), to interdependent (mature adult). There are obvious parallels between these stages and those observed at the level of organizations progressing toward sustainability: inactive, reactive, active, and proactive. In other words, both groups and individuals progress through similar stages. It is therefore straightforward to connect these insights to our previous discussion of the mind-set expansion of leaders moving toward sustainability,

84   Lutz, A., Slagter, H.A., Rawlings, N.B., Francis, A.D., Greischar, L.L., & Davidson, R.J. (2009). Mental training enhances attentional stability: neural and behavioral evidence. *The Journal of Neuroscience*, 29(42), 13418-13427.

85   Goleman, D., McKee, A., & Boyatzis, R. (2002). *Primal Leadership: Leading with Emotional Intelligence*. Boston, MA: Harvard Business Press; Senge, P., Sharmer, O., Jaworski, J., & Flowers, B.S. (2008). *Presence: Human Purpose and the Field of the Future*. Boston, MA: Crown Business.

86   Limb, C.J., & Braun, A.R. (2008). Neural substrates of spontaneous musical performance: an fMRI study of jazz improvisation. *PLoS ONE*, 3(2): e1679. doi:10.1371/journal.pone.0001679.

87   Siegel, D. (2009). *Mindsight: The New Science of Personal Transformation*. New York, NY: Random House.

88   Bakshi, R. (1998). *Bapu Kuti: Journeys in Rediscovery of Gandhi*. New Delhi, India: Penguin Books.

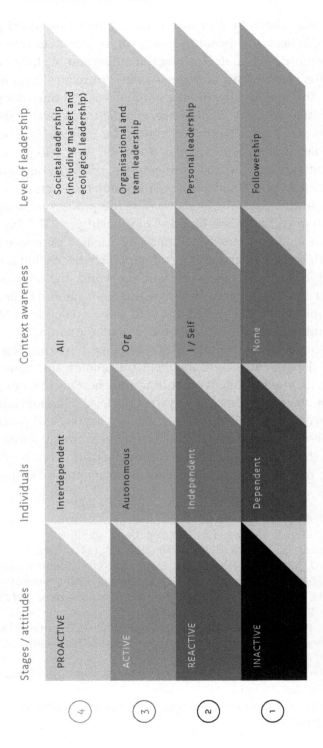

FIGURE 9.5   Stages of leadership development

expressing a broadening of their scope of awareness from personal ("I"), organizational ("Org"), to societal ("All"). These stages can therefore be related to individuals, organizations, levels of awareness, levels of leadership, and leadership styles, as illustrated in Figure 9.5.

The precondition for this progress to happen is that leaders train the regulatory process of their mind, thus expanding their mind-sets and worldviews.

9.4 Integrated leadership models

Returning to the 6C model and the particular mind-sets demanded by societal leadership, it is now possible to develop an integrated model for the creation of sustainable or triple value performance by bringing together the theories of mind-set development, psychological empowerment, self-determination, flow, and transformational leadership, as well as insights from development psychology and neuroscience.

Furthermore, connecting the 6C mind-sets to the familiar questions, "where," "why," "who," "what," "how," and "within," transforms the model into an easy-to-remember management tool. Given someone's "context awareness" (where), the upward line from "centeredness" (why) to "collective vision" (what) corresponds to Cook-Greuter's vertical development—which we define as a process of enhancing "consciousness" (within)—and defines orientation in and nourishment from/for the changing context.

"Creative competence" (how) and "connectedness" (who) correspond to horizontal development; these are necessary leadership skills yet do not necessarily involve the mind-set development that occurs on the centeredness–creative vision trajectory. For success in an organizational setting, attention should be balanced between task *and* people, not one or the other, which is often the case. Tasks need to be coordinated well and relationships need to be productive. Thus, balancing creative competence with connectedness is related to centeredness and creative vision.

When all mind-sets and qualities are dynamically balanced, they generate a strong sense of empowerment and increase the ability to deal with the ever-increasing levels of complexity. This process drives a leader's sustainable performance or capacity for triple value creation. These different levels of leadership can be trained and cultivated such that, at the highest level of complexity, leaders have developed the full capacity to expand their

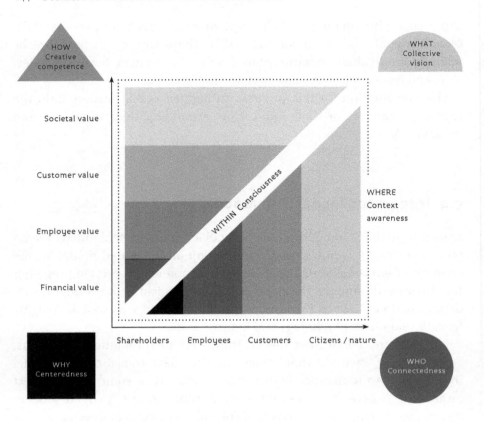

FIGURE 9.6   Societal leadership for triple value creation

mind-set and sense of responsibility to care for themselves, their organizations, and society. This is the level of societal leadership.

We can now bring these insights together with the ideas of value creation for multiple levels of stakeholders, as shown in Figure 9.6. Each type of stakeholder requires a different type of value. At the level of the shareholder ("I") we speak of financial value creation. The next level (moving toward "Org"), employee value is being created. Expanding to the next level ("Market"), the customer becomes an object of (rather than merely a source of) value creation. At the highest level ("All"), societal value is created.

The potential of this model is that it can show leaders a path toward creating sustainable value by, in the words of Peter Block, "a purpose of greatness."[89] It supports the calls of Paul Polman ("what can your company

89   Block, P. (1987). *The Empowered Manager*. San Francisco, CA: Jossey-Bass.

do for society?"),[90] Feike Sijbesma ("taking responsibility for society and nature"),[91] and the Dalai Lama ("when something is of benefit, even in the very long term, then just do it").[92]

On the basis of their context awareness, leaders can develop a collective vision for sustainability which will determine the sense of empowerment that they and their co-workers will experience in pursuing the realization of the vision. Creating a collective vision, setting sustainability goals, and an intrinsic sense of responsibility for the "I," "Org," and "All" levels—in other words, developing a shared purpose toward triple value creation—will lead to a larger degree of intrinsic motivation of their employees than pursuing purely financial goals. The morale and commitment of employees at work will grow as a result of pursuing a genuinely shared collective vision: as employee well-being is strongly correlated to top performance[93] (happy people work more effectively), this process therefore becomes a virtuous circle of both personal and business development. In other words, by creating a linkage between sustainability and employee well-being goals, the bottom line will be enhanced as well. At this point, the value creation capacity of the firm will have reached its zenith: it will create triple value for the organization, for its customers, and for society at large.

If we translate these insights into leadership development programs aimed at creating societal leaders in business, we can create a generation of leaders who embody the new **sustainable–relational economic paradigm**, described at the beginning of this chapter. These leaders can build organizations that are capable of transforming their value chain systems in collaboration with their stakeholders. In this way, companies will be able to become catalysts in reversing the downward spiral of value destruction, which is part of the current economic paradigm, into the upward spiral of shared value creation with society. In other words, the current "competitive race" to the bottom could be transformed into a "compassionate race" to the future.

90   Forum for the Future (2011). Interview with Paul Polman. Retrieved from https://www.forumforthefuture.org/blog/6-ways-unilever-has-achieved-success-through-sustainability-and-how-your-business-can-too.

91   Sijbesma, F. (2013). We need to redesign our economy. *Huffington Post Business*. Retrieved from http://www.huffingtonpost.com/feike-sijbesma/we-need-to-redesign-our-e_b_2597564.html.

92   Paraphrased from comments made in private meeting with the author (see pp. 71-72).

93   Loehr, J., & Schwartz, T. (2004). *The Power of Full Engagement*. New York, NY: Simon & Schuster.

10
Six questions to develop shared purpose

We live in a time when we can no longer ignore or be unaware of the links between global sustainability challenges and individual choices. The issues we face today are unlike any we have encountered, in terms of their scale, impact, and complexity. Yet, as the Dalai Lama has said, it all starts with the individual. Each individual has the capacity to make a personal contribution, which, together, will add up to create positive impact in society.

This book describes how leadership can have a positive impact on societal challenges, through a process of developing "shared purpose." While fully manifesting societal leadership in business is a complex process and a long path, this chapter will help readers to take the first step on this path. There are six clusters of questions, which will help readers to determine what they can do as individuals to develop shared purpose, and progress toward societal leadership. The reader can work on these questions individually or with colleagues and partners.

> The main thing is to have a firm belief about the core responsibility of solving these challenges [yourself], and not delegating that to someone else. If you don't have that inside of you, then we can all sit here and be critical about government and legislation, but where are you in this? You cannot be a bystander in the system that gives you life in the first place … The role of business has to be firmly understood by the CEO down, that it is there to serve the broader society, the common good and only by doing that very well you will be rewarded, but it has to start there and end there.
>
> Paul Polman, CEO, Unilever[1]

There are a number of models presented in this book derived from research from various sources, as we have seen in the previous chapter. In order to test these models in practice, I have applied them to my own process of personal development and growth. The Shared Purpose Model (Figure 10.1) lends itself best as a quick scan for ascertaining one's success in living a "shared purpose." This model can be accessed by reflecting on six clusters of questions that are described at the end of this chapter.

To give readers a sense of what this model can do for you, I will explain how it has guided me in my own life and work. Most significant was the discovery that each time in a certain situation or context (or the "where") when I experienced a sense of "purpose," I was able to know "why" I wanted to do my work, and the impact I could have or the "what" I had to do. At that time I also knew for whose needs I was performing and had access to people who could help me ("who"), and I possessed skills and know-how to do my work ("how"). In other words, the "why," "what," "who," "where," and "how" were in balance in a certain context. This result was a sense of (shared) purpose, where things seemed to happen in flow without excessive effort or worry. This experience would then stretch my consciousness ("within").

The sense of shared purpose that I talk about here is more than the intention, more than the "why." It is also more than the vision, goal, or "what" that you have in mind. It is the result of the "why" and "what" coming together, supported by the other factors of "who" and "how" in a given context or "where." Shared purpose arises out of dynamically balancing these factors.

In my career I experienced a progression of four stages of purpose, which more or less coincided with the timing of the four dialogues with the Dalai Lama. Initially, it took some time to discover my ability to experience purpose (which is **Phase 1: finding purpose—Chapter 1**). In my banking career,

---

1   Confino, J. (2013, October 2). Interview: Unilever's Paul Polman on diversity, purpose and profits. *The Guardian*. Retrieved from http://www.theguardian.com/sustainable-business/unilver-ceo-paul-polman-purpose-profits.

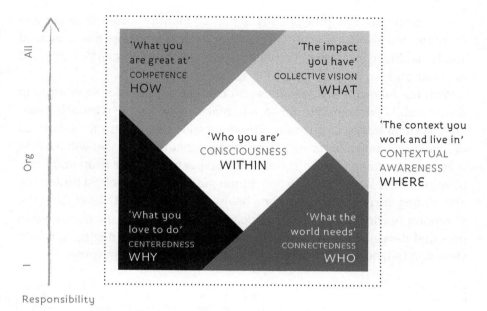

All

Org

—

Responsibility

FIGURE 10.1   Shared purpose model

I had glimpses of it when I was able to use my position and knowledge to help disadvantaged people and communities. But later I lost this ability, as I could not reconcile my intentions with the direction that the bank (and the financial industry in general for that matter) was taking. The "why" and the "what" of my work started to diverge. It took me some time to bring together the wisdom and courage to change the direction of my career. In the years that followed I had a notion of actually living my purpose (**Phase 2: living my purpose—Chapter 3**), although it was still rather unfocused and the end goal was unclear.

After I had chosen to apply my skills to helping business leaders to discover their purpose, I moved to the next phase of finding what I called my "shared purpose" (**Phase 3: finding shared purpose—Chapter 5**). While I enjoyed the broad exploration of the previous phase, I was now clearer about "where" I belonged and therefore I could better focus on "what" I wanted to accomplish. Simultaneously, it became clear with "who" and "how" I could achieve these goals. However, given the continuously shifting context (emergence of the financial crisis and the inability of many companies to embrace the challenge of sustainability), I decided to engage in research, writing, and education, during which phase I discovered a sense of actually living my shared purpose (**Phase 4: living shared purpose—Chapter 7**).

As a result of this process, my sense of identity and self-awareness ("within") has grown exponentially. At each stage, my purpose was enhanced and I was able to improve my capacity to create value. This can be explained in Figure 10.1.

With the following six questions you can get a sense of how to develop your capability of shared purpose, and make a start on the path of becoming a societal leader in business. These questions are meant for you as an individual. To make progress as a team and organization, you will need to reflect on the same questions from the perspective of your team and organization, but this is difficult to do without guidance by a societal leadership and strategy expert. The questions below represent the "I" level from the viewpoint of your personal societal leadership, which anyone interested in personal development can undertake. They indicate the degree to which you as an individual have developed a capacity for shared purpose.

	1 Low	2 Not much	3 Average	4 Much	5 High
Where: Contextual Awareness— "The environment you relate to"					
I have an open and curious mind for what happens in the world around me.					
I can identify the ten key trends and developments in my environment.					
I can identify the impact and opportunities that these trends and developments represent.					
Where: Contextual Awareness— "The environment you relate to"					
I have an open and curious mind for what happens in the world around me.					
I can identify the ten key trends and developments in my environment.					
I can identify the impact and opportunities that these trends and developments represent.					

	1 Low	2 Not much	3 Average	4 Much	5 High
Within: Consciousness—"Who you are"					
I am conscious of my own thoughts, feelings and mind-sets and the impact on my output.					
I have a proactive (learning) mind-set rather than a reactive mind-set.					
How would your best friend score your capacity to be self-aware and manifest a proactive mind-set?					
Why: Centeredness—"What you love to do"					
I love what I do at work, and it gives me meaning and enthusiasm.					
I normally remain calm after confrontations or setbacks.					
I stay true to my values and I know when to reflect on my perspectives.					
What: Collective Vision—"The impact you have"					
I take time to reflect on the impact that my work can have on the world.					
I have a clear future in mind, for the people and the world around me.					
How do you score your capacity for visioning a desired future?					
Who: Connectedness—"What the world needs"					
I am aware of the communities that I and my organization serve, including their needs.					
I can easily see the world from another person's perspective, without bias or judgment.					
How do you score your capacity to connect with different stakeholders at the same time?					

	1 Low	2 Not much	3 Average	4 Much	5 High
How: Creative Competence— "What you are great at"					
I know what I am great at in relation to work.					
I remain resourceful after confrontations or setbacks, because I know where my strength lies.					
How do you score your capacity for achieving results?					
Total					

Now add up your scores and divide the total by 18, which will give you an overall score of the degree to which you have developed shared purpose, as the first step toward societal leadership in business.

0–18   low
18–36   not much
36–54   average
54–72   much
72–90   high

Concluding remarks

Now that the book has been finished, a final word on what I consider to be the main lessons from this lifetime inquiry into the shared purpose of business and economics. It all started with the Dalai Lama's vision for a more compassionate society. In his words, such a society is beyond religious or ethnic conflicts, beyond social and economic inequality, beyond environmental destruction and resource depletion, beyond excessive greed and corruption—a world where people recognize their common humanity and where kindness and well-being for all has become the norm for all social, political, and economic interaction. For the Dalai Lama, it is not a question of "if" this vision materializes into reality, but "when." It is this conviction, which he so naturally embodies, that makes his vision for the future so inspirational.

How can business contribute to the realization of this vision? This question has now guided me for many years. The question turned into a quest—with many ups, downs, and unexpected twists along the way. It has been a fascinating journey through the worlds of law, economics, business, finance, science, and personal development. Clearly you could look at business in two ways: either as part of the problem or as part of the solution. I chose the latter. It was clear to me that it would be impossible to solve the global challenges without using the unique institution of business and stimulating its powerful creative force. Only business can create prosperity, but the question has become: prosperity for whom and for what purpose?

I have seen that an increasing number of companies are actively searching for solutions, not because they are idealistic but because they are discovering that, in today's interconnected world, this is in their enlightened self-interest. By taking measures to adopt more sustainable practices, these companies are creating the conditions to thrive in the mid- and long term. Such companies are not only taking actions to future-proof themselves, but are also smartly leveraging a strong zeitgeist of purpose-driven behavior in society in general. Groups of health-conscious consumers, meaning- and balance-seeking employees, eco-conscious regulators, future-focused students, and increasingly influential value-based communities are influencing business. Low costs and open-source online platforms allow these groups to engage with business directly and become a creative force for societal change. They generate hope that capitalism will be able to reinvent itself. At the very least, they are manifestations of a **new paradigm of interconnectedness** in business.

My journey through modern scientific development revealed a similar shift in paradigm. Both mainstream economics and business still speak of scarcity, market efficiency, and material growth. This is rooted in the paradigm of the so-called *Homo economicus* motivated by power, greed, and selfishness; in short, the neo-Darwinian model of survival of the fittest. However, current neuroscience indicates that it is contradicted at the level of our existence as biological beings, where research has shown that we are at least equally driven by motives of care, altruism, and responsibility. As "social animals" we are deeply dependent on our social and ecological environment and we have an innate sense of fairness and care for those around us. In a sense, we still live with a limited and "outdated" understanding of ourselves, our motives, and our potential. With the revelatory insights from new branches of scientific research, we can remove this misconception and discover more fully who we are and what we can become as humans and society.

What do these findings mean for business? I can summarize in four main points:

1. The world has become increasingly **interconnected**, as evidenced by the unprecedented societal, economic, and ecological challenges that we face. Business can no longer run on the principles of "business as usual." As succinctly stated by Feike Sijbesma, CEO of DSM: "How can you be successful as a business in a society that fails?"[1] Obviously, you cannot. The new "business of usual," therefore, is sustainable business for both economic and societal progress.

[1] WBCSD (2012). *Measuring Impact Framework: A Guide for Business*. Geneva: World Business Council for Sustainable Development.

2. The purpose of business, therefore, is to create sustainable value. The most sustainable value that business can create is value that does not distract from, but rather contributes to, the long-term prosperity and well-being of society and ecosystems—because that is what, in the long run, business depends on. I have called this **triple value**—value for one's organization, customers, and society. This goes beyond CSR. In contrast to CSR, which is often at the periphery of business, triple value creation is at the core of business, including strategy, sales, marketing, finance, and operations. Ultimately, behind every market demand there is a certain need in society. If one cannot find that need, one's business is not creating triple value. Unfortunately, this logic is not yet reflected in mainstream indicators of business performance, such as the profit and loss statement, share price, market share, or, at a national level, gross domestic product, each of which captures only financial value. This means that we have to develop **comprehensive triple value indicators**.

3. Creating triple value in business requires a **new compass for leadership**. This compass will help leaders to identify the societal needs that their business can serve. This is a process of finding one's purpose. The most powerful purpose is a **shared purpose** between business and society. Without such shared purpose, employees will suffer from a "purpose gap"; they will feel disempowered to give their full selves to the organization. Shared purpose, therefore, is a natural ally of sustainability—at a personal and an organizational level, as well as at a societal level. It is a key subset of sustainability, a must-have component for the true future flourishing of all employees, present and future. This is highly relevant to human resource professionals in their role as recruiters, leadership developers, and coaches. Finding one's shared purpose with society and anchoring the organization accordingly will be the prime role of the business leaders of the future.

4. For leaders to succeed in this role, they will need to **develop and expand their mind-sets and worldviews**. Not only are mainstream business indicators misleading, the underlying assumptions that business leaders hold are no longer adequate for the complex world of today. Through exploration of our deeper human nature, we will discover new ways of leading organizations. The 6C model

presented in this book will help leaders to retrain their minds and to develop the competences for **societal business leadership**. Through this enhanced and updated form of leadership, leaders will become more effective in business and create well-being for themselves, their organizations and the world.

What is next? There is still a long way to go. The logic of sustainable business and triple value is still poorly understood, yet time is running out to prevent further decline and even the collapse of our political, economic, and ecological systems. Vested interest, perverse incentives, and populist politics stand in the way. Sometimes I feel impatient and discouraged. At these moments I reflect on the words that the Dalai Lama shared at the first public dialogue in 1999:

> My own experience has been that no matter how numerous the difficulties and no matter how big the obstacles, if your belief or your ideal is truly reasonable and beneficial, then you must keep your determination and maintain a constant effort. I also think that if something is right and good for the larger community, then whether that goal materializes within one's own lifetime or not doesn't matter. Even if it's not going to materialize within our own lifetime, we have to keep working. The next generation will follow and, with time, things can change.

These words renew my energy and commitment. They help me to realize that I have just one small part to play and I can trust that others will build on what my partners and I have done. I hope that this book will encourage others, collectively, to help to turn business and economics into instruments of positive societal change.

Positive capitalism and the opportunity of mirror flourishing: the grammar of interconnection is the business discipline of our age

Afterword by David Cooperrider

Our generation has arrived at the threshold of a new era in human history: the birth of global community. Modern communications, trade and international relations as well as the security and environmental dilemmas we all face make us increasingly interdependent. No one can live in isolation. Thus, whether we like it or not, our vast and diverse human family must finally learn to live together. Individually and collectively we must assume a greater sense of universal responsibility ...

You see, positive things do not come by nature. For positive things we have to make an effort. We must make the effort. Nobody, no one else, can do that. So everyone, hope for a better future, a happier future, if that is our wish. The present generation must make every effort. It is our responsibility.

The Dalai Lama[1]

1 Address to the Members of the United States Congress in the Rotunda of the Capital Hill in Washington, D.C. by His Holiness the Dalai Lama, April 18, 1991. Retrieved from http://www.oocities.org/~spiritwalk/dlamagovt.htm.

As someone who has spent more than a decade and a half watching the work of Sander Tideman and his colleagues unfold[2]—learning from him, marveling at his synthetic insights, and collaborating with him in business and society dialogues with His Holiness the Dalai Lama in concert with extraordinary CEOs from Medtronic, Calvert Funds, Green Mountain Coffee Roasters and many others—I am thrilled with this inspiring, original, and far-sighted business book. Today when I speak with senior executives as well as with young up-and-coming management students about the future of leadership, I guide them to three books: Gary Hamel's *The Future of Management*;[3] Chris Laszlo and Judy Brown's *The Flourishing Enterprise*;[4] and now, this dazzling work *Business as an Instrument for Societal Change: In Conversation with the Dalai Lama*.

This is what is so stunning about this book: it's not just about the future, but it will actually help shape it. Much like Canada's legendary hockey superstar Wayne Gretzky who famously declared, "I never play the puck where it is but where it is going to be," this book opens the doorways of the mind, our

2   I want to thank Anders Ferguson for his vision and for bringing me together with Sander Tideman and other remarkable colleagues such as Marcello Palazzi, for that early global conference in New York on "Spirit in Business" (SiB) in April 2002 (described in this book). It was a watershed conference that brought together 600 business and society leaders, and in many ways was a catalytic moment not only in the development of this story, but also for a massive wave of spin-offs. Soon after facilitating that design-inspired SiB conference, I was next invited to bring our Appreciative Inquiry Summit method and the theme of "business as an agent of world benefit" to the United Nations where, together with Nobel Laureate and Secretary General of the United Nations, Kofi Annan, we helped create the growth strategy for the UN Global Compact. Today over 8,000 corporations participate and 500 business schools have come together to advance many of the ideas articulated so powerfully in this book. And at my own school, Case Western Reserve University, an academic center to advance cutting-edge science in this area was soon established in the form of the Fowler Center for Business as an Agent of World Benefit. I am forever indebted to the courage, vision, and wisdom of this team, and I offer this afterword as a small token of my deep gratitude. This team, together with this book, is a gift. This work, to be sure, has given me more hope about our world's future than anything I've ever been involved with, and so much of it began at that fertile verge called SiB.

3   Hamel, G., & Breen, B. (2007). *The Future of Management*. Cambridge, MA: Harvard Business School Publishing.

4   Laszlo, C., Brown, J.S., Ehrenfeld, J.R., Gorham, M., Pose, I.B., Robson, L., ... Werder, P. (2014). *Flourishing Enterprise: The New Spirit of Business*. Stanford, CA: Stanford University Press.

world-view, *and* the concrete practices for winning the hearts and minds of customers and communities, employees and partners, and society at large *while* articulating the next episode in capitalism as a whole. To be sure this is a business book, it's about leadership that unites success and significance into one integral whole, and it's about flourishing enterprise.

The key concept is **shared purpose**, which ignites new sources of relationship value—called **triple value**—and it is a positive sum economic, human, and ecologic trajectory that "is in the cards" so to speak, because it is based on the fundamental reality of interconnection.

Sometimes it takes the unified insight of many fields at once to achieve a massive shift, in this case toward a consciousness of connection. And that's precisely what this book does. It achieves an enormous resonance by uniting the scientific world-views of cutting-edge physics, neuroscience, and the positive psychology of human relationships (we are hardwired for compassion) together with ancient spiritual wisdom and the kinds of reflective practice that speak to the oneness of humankind and all of life. The result: the dramatic and totally practical realization that business can be one of the most positive forces on the planet and that we can, right now at this time, create a convergence zone where businesses can excel, humankind can flourish, and nature can thrive.

> We can, right now at this time, create a convergence zone where businesses can excel, humankind can flourish, and nature can thrive.

Moreover this volume is not just theoretical. Some of the most magnetic pieces of writing in this book, as the reader has no doubt experienced, is Sander Tideman's own growth story, and his penetrating questioning and transformation as a senior executive banker at ABN AMRO Bank. And precisely because this is written by a successful executive, the research findings, instead of being dry and inert, are presented here with delicate precision and joyful clarity—and just enough for those who want the powerful business case, the theories and data on human flourishing, the leadership case for triple value, and the case for positive change. The personal storytelling of the author, for example, brings everything into special focus because the storytelling is honest, heartfelt, and humble. You cannot help but reflect on your own life as the author narrates his own life-enriching, life-expanding transformation. And I found myself vicariously inspired. We thrill to people who have the courage to open themselves to the vertigo of new vision—and the letting go of the cultural blinders of our times—only to find a prize in the waiting. In reading the book I was reminded of one of Mark Twain's precious insights when he said: "The two most important days of your life are the day you were born and the day you found out why."

In this case I doubt if the author, in his earlier years as a lawyer and banker, ever imagined himself designing some of the most enlightening, rich, and historic dialogues on the planet with world-respected CEOs, such as Bill George of Medtronic (now CEO in residence at Harvard), *and* His Holiness the Dalai Lama. I participated in those privileged dialogues, and everyone, myself included, sat on the edge of their chairs.

The grammar of interconnection is the business discipline of our age

In the rest of this afterword I would like, almost like a jazz musician, to play off some of the notes sounded across the four dialogues. I want to zero in on how participating in the early dialogues described here inspired me personally (thank you Sander Tideman, Anders Ferguson, and Marcello Palazzi) and called on my colleagues and I to design and scale up the most important research project of my career. Ever since we created the interview guide we created following the Spirit in Business conference (described in Chapter 3) we have to date, together with thought leaders such as Peter Senge of MIT and Jane Nelson at the Kennedy School of Leadership at Harvard, completed over 3,000 Appreciative Inquiry interviews into the study of "business as an agent of world benefit." It is a systematic, global study that's given me more hope about our world's future than anything I've ever worked on. This "World Inquiry," as we called it, lifts up thousands of innovations from every continent and culture: business as a force for peace in high-conflict zones; business as a force for eradicating extreme poverty; and business as a force for the epic transition to a bright green economic and renewable-energy future—all of this while elevating, expanding, and extending the success of the business.[5]

What I've come away from in our studies is one huge conclusion and one equally huge confusion. The conclusion: it's that the idea of "business as an agent of world benefit" is the business opportunity of the 21st century and represents the catalytic frontier in innovation. We are on the eve of one of

5   For many more resources visit the "AIM2Flourish" section of the Fowler Center for Business as an Agent of World Benefit website at https://weatherhead.case.edu/centers/fowler/.

the greatest revolutions in management history, an era of deep-seated transformation, where high purpose and the pursuit of societal leadership is emerging as the most powerful unifying thread for propelling industry-leading innovation in complete and simultaneous convergence with solutions to the call of our times. In study after study the industry leaders we've been tracking—e.g., Unilever, Whole Foods, Tesla, Interface, Tata, Kingfisher, and many of the well-known "firms of endearment"—are all moving toward Sander Tideman's concept of "triple value" which involves creating shared value for shareholders and stakeholders including a flourishing biosphere. Moreover, as these industry leaders experience increased business success (turning waste to wealth; more engaged and passionate people; less risk and litigation; more trust and higher-quality connections across the supply chain; dazzling and out-of-the-box design thinking; the opening of whole new sustainability-inspired markets; factory designs that produce more clean energy than they use; industry-wide positive disruptor capability; more customer delight; combining success with significance, etc.) they are also rapidly advancing from a view of sustainability-as-less-harm to a view of sustainability-as-flourishing—as net positive, not just net zero impact, as a regenerative force for good.

> They are rapidly advancing from a view of sustainability-as-less-harm to a view of sustainability-as-flourishing—as net positive, not just net zero impact, as a regenerative force for good.

In today's increasingly dynamic, interdependent, and unpredictable world, the most successful businesses realize that to create long-term value for investors they must work together with key partners, communities, and stakeholders in ways that are mutually beneficial, not only as a relational responsibility but as a gold mine for innovation.

That's what happened to Tesla when it set out to respond to the future. It found that, by listening to the younger generation and playing the puck where it's going to be, it was not even an auto company. Elon Musk, not right away but after much reflection, came to the conclusion that "the world does not need yet another auto company and our car is not that important ... what the world needs is a sustainable energy revolution,"[6] and that's

---

6   Mace, M. (2016, May 5). Elon Musk: "We need a revolt against the fossil fuel industry." *The Guardian*. Retrieved from https://www.theguardian.com/environment/2016/may/05/elon-musk-we-need-a-revolt-against-the-fossil-fuel-industry.

what's calling Tesla partners to innovate every day. When we did interviews globally at Tesla stores everywhere from Amsterdam to Columbus, Ohio, what do you think we found? We found Tesla's associates beaming with energy, insight, and delight—and teamwork, lots of teamwork. The phrase "employee engagement" sounded so old after our interviews. What we experienced were people on a mission, the kind of shared purpose and intrinsically motivated mission Sander believes will fuel great sustainability breakthroughs. The people we met loved their work, used the word love over and over, and loved thinking beyond the possible *for the world.*

So the conclusion that this volume asserts is that all of this is a tremendous trajectory, not just a fad or a minor trend. But the confusion I have is this: where should companies new to all this actually begin and start their journey? And why—as observers of the field have begun to document— have so many sustainability initiatives run out of steam? From BP's dislodgement of its original sustainability vision ("Beyond Petroleum") to Green Mountain Coffee Roaster's fall from the sustainability leadership charts (once its visionary CEO Bob Stiller retired), and then from Walmart's "all hands on deck" early big splash to Interface's Mount Sustainability uphill battle, there appears

> Where should companies new to all this actually begin and start their journey? And why have so many sustainability initiatives run out of steam?

to be a brick wall. "Climbing Mt. Sustainability gets increasingly harder the closer you get to the peak," said the legendary CEO Ray Anderson, "just as climbers of Mt. McKinley show us as they are gasping mightily for air the higher they go."[7]

For the first time, after reading and rereading the four CEO/Dalai Lama dialogues, and after reading Sander's description of the Bodhisattva, I feel like I have an answer. And as I shall explore, the journey to "doing good and doing well" might be simpler than we all think if we:

- **Take seriously the new view of the human being.** The combined view from biology and neuroscience that the average person is guided by an innate sense of fairness and care, and that altruistic motives are inherent in the very first interactions of a mother and child; and, as the Dalai Lama suggests, we actually need to practice

7   Ray C. Anderson Foundation (n.d.). *Climbing Mount Sustainability.* Retrieved from http://www.raycandersonfoundation.org/assets/pdfs/rayslife/Essay ClimbingMountSustanability.pdf.

compassion for others "because it benefits us" and "the more compassionate one's mind, the happier one feels." This is not to say that *Homo economicus* is not real, but it is to say, like Adam Smith wrote in *The Theory of Moral Sentiments*, that ethics (doing things right) and economics (creating value) must be fundamentally tied together and that, when they are, there is a mutually elevating bidirectionality that is natural and powerful: doing things that are right and beneficial "out there" brings benefit "in here." It involves what the Dalai Lama terms "a wise selfishness." Later I will talk about and illustrate this from our own studies on business as an agent of world benefit, as "mirror flourishing."

- **Take seriously the shift from the global and "social responsibility" of business to the global and "social intimacy" of business.** This is a huge shift. It's implied in this volume at every turn, from the spiritual perspective that we are all one, to the recognition in quantum physics postulated in string theory which implies that everything—visible and invisible, material and mental, close and distant—is connected and entangled. For me the business implications became totally clear, and relevant, through meeting CEOs in Brazil. One commented that the language of corporate responsibility—just the language itself of one entity taking responsibility for the whole—felt like an ethical demand or another external "have to." That CEO was Rodrigo Loures, one of the most successful and respected business leaders in the country, and he said: "it's not about *responsibility for* the whole; it's about *intimacy with* the whole." Read Rodrigo's words once again; they are that important. What a powerful, quiet recognition. It changes everything. The shift from responsibility to intimacy does not change the scene as much as it changes perception: we don't wish for our son or daughter or other cherished, close relations to simply survive; we want them to thrive. We want them to flourish and, when they do, we thrill to their success—*because we care*—it's a response that is intrinsic. So can we expect the same natural conviction or instinctive response from our societal leadership initiatives? Where can we look for the *intrinsic* energy and natural motivation to drive large-scale change for sustainable business, whole industries, and sustainable economies? Are there ways to access whole new magnitudes of change capacity?

- **Take seriously the fact that all system change (not only technical or structural change) ultimately requires leadership.** The opportunity to embrace triple-value contribution can't just be an external transformation or the mission statement on the wall. It requires working on the inner self/mind: it requires unlocking a human being's full potential, all the qualities or strengths of character that we are capable of developing, for the benefit of transforming larger and perhaps wider circles of system change, toward mutually beneficial outcomes.

A recentering on the human spirit

What I admire most about the shared purpose concept, so aptly articulated by Sander Tideman, is its courage to take up the "spiritual adventure" behind sustainability-as-flourishing, the search for meaning, purpose, and value that becomes an end in itself. As I have documented elsewhere, one of the truths of our time is this hunger deep in people all across the economy for realizing that their lives count, and count affirmatively as it relates the greatest challenges we all share. Is it an accident that young people are flocking to a company such as Google that provides dedicated programs for personal development, wellness, and nutritional excellence, hiking trails and solar-inspired buildings, meditation courses, and collaborative teaming, as well as daring, world-changing projects such as the one to bring the internet to billions for free—all the world's information and instantaneous connectivity—in order to eradicate extreme poverty within a generation?

When you visit the Googleplex, as I just did last year, you see extraordinary investment in the inner dimensions of human consciousness. Do you know what is perhaps the most popular course by demand at Google? It's the one offered by Chade-Meng Tan, employee number 107, and it's based on his book *Search Inside Yourself: The Unexpected Path to Achieving Success, Happiness (and World Peace).*[8] In terms of numbers, this course is rivaled only by the likes of Harvard's most popular course ever, a course on the positive psychology of the good life. Talk about success *by* design. While most companies espouse that people are the most important element in their success, here one could sense the genuine commitment being acted

8   Tan, C.-M. (2012). *Search Inside Yourself: The Unexpected Path to Achieving Success, Happiness (and World Peace)*. New York, NY: HarperOne.

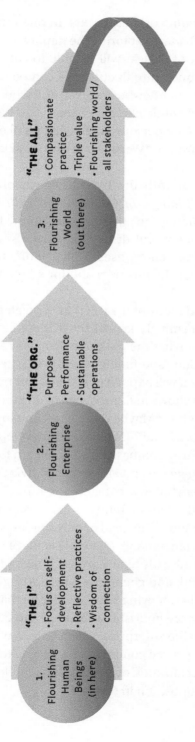

FIGURE 11.1   Toward shared purpose, triple value, and business societal leadership for good
What if we reverse this sensible sequence and start at #3 and paradoxically enable #1?

on in every nook and cranny of the campus. In one meeting, I watched a team that was on fire with ideas exploring the implications of a world where sensors via "the internet of things" will be able to cut world energy costs in half, where ocean temperature fluctuations can be monitored just as a mother closely attunes to her newborn's fever, and where global vital signs of species, toxicities, and subtle climate changes can be sensitively examined 24-7—everywhere. One person shared what it's like to work on such a world-changing legacy project. "We are," she said, "waking up to our earth's vital signs."

All of this is consistent with the hypothesis explored *Business as an Instrument for Societal Change: In Conversation with the Dalai Lama*—that organizations building on the human spirit for flourishing in terms of inner development will drive more successful organizations as "sustainability+flourishing" enterprises, which ultimately will drive sustainable value creation for the business and for a better, more flourishing world (see Fig. 11.1.)

There is, as Sander and many of the CEOs in the dialogues with the Dalai Lama demonstrate, a seemingly logical set of stepping-stones here. And this sequence of stages—what the author articulates as a focus on "the I" and then "the organization" and finally "the all"—invites a whole series of important questions: what will happen to the quest for triple value and societal business leadership efforts in the absence of an inner flourishing and a consciousness of connection? Maybe that's why so many early sustainability leaders have hit the brick wall? Can we really develop human beings for character development (e.g., compassion and concern for humanity, empathy, wisdom, social intelligence and altruism, creativity, ethical affirmation of life, mindfulness, generativity, concern for future generations, etc.)? And then, to play with the ideas a bit, we might ask: can the trajectory in Figure 11.1 work bidirectionally instead of only unidirectionally? What if we reversed this sensible arrangement? Might we achieve the prize of a flourishing inner life—inspired with high purpose, innovative, engaged, compassionate, more mindful, and generative—by starting first with the third circle on the right and then moving left? Could it be that the effort toward, and active creation of, more flourishing societies and ecosystems awakens more interest in sustainable enterprise, which in turn works through backward causation to create fertile conditions for people to seek out reflective practices leading to the realization of an inner sense of flourishing—in other words, traversing the proposition in reverse?

We know from systems theory that, when any two events are related inter-dependently, describing one as cause and the other as effect is an arbitrary designation. Both Peter Senge[9] and Karl Weick[10] have shown that, in causal loops, no variable is any more or less important than any other variable. In a loop you can, at least theoretically, start the sequence anywhere you want by changing any event or variable.

There can be no flourishing "in here" without flourishing "out there"

I would like to add at least one additional dimension for leveraging Sander's model and for leaders to consider in terms of an action agenda or positive pathway. For me the implications of looking at this book's compelling prop-osition in reverse order is not only interesting, but also powerfully action-able as a leverage point for one of the most remarkable sources of business and human value that will never run out. It has to do with self-amplifying loops and virtuous circles. "Managers get in trouble" argues Karl Weick, "because they forget to think in circles."[11]

One of the high-point moments in recent memory was a series of almost a dozen lectures I did in duet fashion with Marty Seligman across Australia. Hosted by the professional services firm PwC, we spoke to hundreds of exec-utives in the financial, healthcare, manufacturing, education, and informa-tion technology fields. Marty is well known around the world as the father of the "positive psychology" movement, which is all about the scientific study of the good life—what is it, where is it happening, and what nurtures it—including the character strengths that enable individuals and communities to thrive. The field is founded on the belief that people want to lead mean-ingful and fulfilling lives, to cultivate what is best within themselves and others, and to enhance their experiences of love, work, and play. Positive psychology, at the early stages, set forward three central pillars of concern: the study of positive emotions, the identification of positive individual

9   Senge, P.M. (1990). *The Fifth Discipline: The Art and Practice of the Learning Organization*. New York, NY: Doubleday Currency.

10 Weick, K. (1979). *The Social Psychology of Organizing*. New York, NY: Addison-Wesley.

11 Weick, K. (1979). *The Social Psychology of Organizing*. New York, NY: Addison-Wesley.

traits or strengths, and the discovery and design of positive institutions.[12] Understanding positive emotions entails the study of contentment with the past, flourishing in the present, and hope and optimism for the future.[13] Understanding positive individual traits consists of the cataloging and study of our highest human strengths and virtues, such as the capacity for love, courage, ethical compassion, resilience, creativity, curiosity, integrity, self-knowledge, justice, spirituality, and wisdom.[14] Understanding positive institutions, as my colleagues and I have defined it, entails the study of how organizations and communities themselves can become vehicles for the elevation, magnification, and refraction of our highest human strengths *beyond the organization* and out into the world.[15]

In our talks, Seligman shared a preliminary outline of what would become his next major book, *Flourish: A Visionary New Understanding of Well-Being*,[16] and I shared the theory of how we become what we study, that is how our appreciative inquiries into the true, the better, and the possible actually create a momentum and new language of life for scientific co-construction of social reality.

We explored the dimensions of flourishing and Marty shared the well-researched dimensions of the good life—five pillars of the flourishing life—through the acronym PERMA (see Sander's discussion of this on page 77). In many respects the PERMA model represents a great summation of the extraordinary findings of positive psychology from the last decade. "P" stands for the study of positive emotion and explores questions such as: "What good are positive emotions such as hope, inspiration, and joy?" "E" signifies "the engaged life," or a life where our unique human strengths are engaged, and how this pillar of wellbeing and growth is actively applied to the workplace. "R" underscores how much flourishing depends on high-quality "life-giving" relationships and the centrality of the Other in a theory of flourishing. "M" is all about the role of meaning—and how, without a life

12   Seligman, M., & Csikszentmihalyi, M. (2000). Positive psychology: an introduction. *American Psychologist*, 55, 5-14.
13   Fredrickson, B.L. (2003). The value of positive emotions. *American Scientist*, 91, 330-335.
14   Peterson, C., & Seligman, M. (2004). *Character Strengths and Virtues: A Handbook and Classification*. Washington, DC: APA Press and Oxford University Press.
15   Cooperrider, D., & Godwin, L. (2011). Positive organization development. In K. Cameron & G. Spreitzer (Eds.), *The Oxford Handbook of Positive Organizational Scholarship* (pp. 737-750). Oxford, UK: Oxford University Press.
16   Seligman, M. (2010). *Flourish: A Visionary New Understanding of Well-Being*. New York, NY: Free Press.

of meaning and purpose, there can be no deep sense of flourishing. And finally "A," or accomplishment, is about the part of wellbeing that is not fleeting but enduring.

Following PERMA's introduction, it was my opportunity to explore not just the individual psychology but also the opportunity for institutions. My call was to share observations on the most flourishing workplaces I had ever seen over some 30 years in the field of management. What surprised even me was this: all of the most extraordinary examples I spotlighted were from organizations leading the way in the sustainable value domain. In the video clips—scenes from our large group Appreciative Inquiry summits with organizations such as Fairmount Minerals (which soon came to be an industry-leading star financially and was also awarded the #1 corporate citizen in the U.S.A.); or scenes from our work with Kofi Annan and 500 CEOs designing the strategies for the UN Global Compact; or scenes from our whole system-in-the-room Appreciative Inquiry summits with cities such as Cleveland's "a green city on a blue lake" work; or our strategic planning summits with whole states such as the work with Massachusetts Governor Deval Patrick's convening with National Grid of more than 300 energy organizations to design the pathway to renewable energy transformation—in each case, what we tracked was a remarkable rise on each of the dimensions of PERMA's flourishing. So what exactly was happening here? As people worked together on new designs for "out there"—for example, the regeneration of a waterway back to its purest potentials and better—there was, in palpable terms, a PERMA response, and the thought crossed my mind: "What do we know about the enhancement of human capacities through the power of restoring and revitalizing nature? Can companies, by engaging people in radically reducing energy watts, also in a reverse fashion actually produce more human energy?"

After a couple of talks I realized it was not a minor discovery or finding. For over 30 years I have been active in the applied side of organization change theory, helping to guide strategic planning and major organization development initiatives in organizations such as National Grid, Apple, Sherwin Williams, Clarke, VitaMix, Google, and the UN Global Compact including companies such as Novo Nordisk, Telefónica, and Tata Industries.[17]

> All of the most extraordinary examples I spotlighted were from organizations leading the way in the sustainable value domain.

[17]   Cooperrider, D. (2012). The concentration effect of strengths. *Organizational Dynamics*, 42(2), 21-32.

Obviously over the decades in the field of management, I have seen a myriad of developments: the birth of the World Wide Web, reengineering of the corporation, participative management, the quality revolution, and many more. Because of my organizational science background I've also had a keen interest in how each particular management innovation affected the human side of enterprise—things such as inspiration and hope, engagement, entrepreneurship and innovation, and the building of emotional intelligence and supercooperative capacity. And herein lies my #1 observation, after 30 years from the real world: there is nothing that brings out the best in human enterprise faster, more consistently, or powerfully than calling the whole organization to design net positive sustainability innovations to humanity's greatest challenges.

> There is nothing that brings out the best in human enterprise faster, more consistently, or powerfully than calling the whole organization to design net positive sustainability innovations to humanity's greatest challenges.

As soon as people come together to accomplish "doing good" out there—that is, concentrating and connecting their strengths in the service of building a better organization, or city, or world, they too begin to activate the PERMA mechanisms for their own and others' flourishing.

Put another way, the active pursuit of sustainable value creation through Sander's concept of societal leadership is not only about serving or satisfying external stakeholders, it is also core—I would now say indispensable—to individual flourishing and flowering inside the firm. Sustainable value creation and shared wellbeing reinforce one another and thereby serve to raise a far-reaching exploration: what is the link between advancing sustainability for a flourishing Earth, with the flourishing of the human side of enterprise? How, precisely, might an organization's quest for sustainable value bring out the best not just on the outside—helping to advance a better society or world—but also almost simultaneously on the "inside"—in the flourishing of more compassionate people, the quality and wisdom of their relationships, their health, their performance, and their capacity for growth, resilience, and positive change?

To be sure, this is not simply a theoretical question. If consistently true, then this is big news for the human resources industry and for every single leader who would love to have a workplace that is alive with purpose, meaning, passion, and high engagement, and overflowing with motivation, trust, inspired innovation, and supercooperative agility. Could it be that the quest for sustainable value based on Sander's proposal around

shared purpose—when everyone in the corporation is galvanized in a coherent way for strategic
sustainable innovation in the service of a more
flourishing world—is the most significant human
development opportunity of the 21st century?

> Could it be that the
> quest for sustainable
> value based on shared
> purpose is the most
> significant human
> development opportunity
> of the 21st century?

In a moment I want to talk about the big insight
Sander Tideman brings to this discussion—it was
a major "aha" moment for me—when he details
the Bodhisattva who does not make a distinction
between self and others. But for now I simply want
to share a nuanced understanding of the paradox of spiritual practice: do
we advance most when we focus on our own inner life (such as the sense
of wellbeing and even bliss we experience in meditation), or when we bring
that spirit of consciousness into life via action, so that we can touch life,
improve the world, and benefit life and others? In the epigraph of this chapter, and in many of the dialogues, the Dalai Lama emphasized the power of
action. He said:

> You see, positive things do not come by nature. For positive things we
> have to make an effort. We must make the effort. Nobody, no one else,
> can do that. So everyone, hope for a better future, a happier future, if
> that is our wish. The present generation must make every effort. It is
> our responsibility.

And this emphasis on action and effort as a vehicle for human development, for me, had echoes with another well-known quotation and theory
of change, attributed to Aristotle: "We do not act rightly because we have
virtue or excellence, but we rather have those because we have acted rightly.
We are what we repeatedly do. Excellence, then, is not an act but a habit."

The power of mirror flourishing

Elsewhere I have speculated further, with my colleague Ron Fry, on this
linkage between sustainable value and PERMA. We've called it the "mirror
flourishing" effect.[18] We could have labeled it many other things and we considered them all—reverse flourishing, positive transference, the so-called

---

18   Cooperrider, D., & Fry, R. (2012). Mirror flourishing and the positive psychology
     of sustainability. *Journal of Corporate Citizenship*, 46, 3-12.

"helper's high," reflexive flourishing, or the hypothesis on "why good things happen to good people."[19]

But the word "mirror" seemed to offer the conceptual richness we were looking for. In neuroscience, for example exploring the relationship between connections and contagion, there has been the conceptualization of a biological basis for empathy, the spread of emotion, and interaction consonance. It's called the "mirror neuron" system, where certain parts of the brain light up when we merely observe a tennis match—just *as if* we were ourselves actually playing the match.[20]

The discovery of the mirror neuron as a concept is shaking up numerous scientific disciplines, shifting the understanding of culture, empathy, philosophy, language, imitation, and the spread of happiness across networks in a synchronized or consonant way. The concept of the mirror neuron helps explain the dynamic of consonance across living systems, the property of being alike, in harmony with, becoming at one with, or a growing together. Of course this growing together can work for good and ill. When our companies are involved in destroying nature or value in the world—think for example of how the despairing and disheartened people of BP were feeling in relation to the horrifying experience and horrifying images of the Gulf oil spill—when it would not stop the spewing of millions upon millions of gallons of oil into the once-blue ocean. I was there so it is easy for me to sense how the human side of that enterprise might enter a state of dissonant discontent or languishing, the very opposite of flourishing. There are colossal human costs of being part of destroying value, and much of the heart-sickness you see in our world today happens because we know, deep down, that environmental and economic collapse is not separate from our lives.

Mirror flourishing suggests an intimacy of relations between entities to the point where we can posit that there is no outside and

> Mirror flourishing is the concrescent flourishing or growing together that happens naturally and reciprocally to us when we actively engage or witness in the acts that help Nature flourish, others flourish, or the world as a whole to flourish.

19 See the comprehensive review in Post, S. (2007). *Why Good Things Happen to Good People.* New York, NY: Random House, Inc.

20 Christakis, N., & Fowler, J. (2009). *Connected: How Your Friends' Friends' Friends Affect Everything You Feel, Think, and Do.* New York, NY: Little, Brown, & Company.

inside, only the creative unfolding of an entire field of relations or connections. The Dalai Lama repeatedly spoke in the dialogues about the reality of interdependence and of understanding how the consequences of our actions manifest in the longer term. This is an echo of Martin Buber who, in 1937, wrote, "In the beginning is the relationship."[21] In a similar manner the metaphor of the mirror neuron helps us erase the traditional boundaries of separation. I will define mirror flourishing as the concrescent flourishing or growing together that happens naturally and reciprocally to us when we actively engage or witness in the acts that help Nature flourish, others flourish, or the world as a whole to flourish.

I think of the smiles I see when I speak to sustainability entrepreneurs. One of those companies is a Dutch shoe-manufacturing company called OAT, who asked their designers: "Can you envision and develop a gym shoe that young people will love, totally biodegradable, and that help regenerate the planet instead of promoting landfill?" Well they did it. When your shoes are worn out you plant them in the backyard and they turn into a flower. When I spoke to the OAT entrepreneurs who created a business for "shoes that bloom," I experienced a sense a joy and a delight in their words that was contagious. I too smiled when I wore my first pair of those sneakers. And then the emotion spread: it was repeated all over again when my children opened their presents, and then again when they asked if I could get the shoes as a gift for their friends. Does flourishing via the waypower of sustainably significant action flow through networks, just as a virus might? Perhaps. In fact the sociology of networks shows that, when a friend living less than a mile away becomes happy, it can increase the probability that you are happy by 25% or more—in other words, our emotions and states of wellbeing, even dimensions of our physical health, flow quietly through our connections. Social and ecological networks, including those with the more-than-human world, are sensitive, intricate, and perhaps even hardwired.

Mirror flourishing, then, is a more than a tangential episode; it is a predictable and observable trajectory. It is, in a word, a huge developmental force if we know how to harness it: we *can* intentionally and consciously create a flourishing workplace by extending beyond ourselves, by immediately working to build a better world that flourishes. We gain life inside by nurturing life outside. And this, as we shall see, is a testable hypothesis: people that experience themselves, their organizations, and their relations

21   Buber, M. (1958). *I and Thou* (New York: Scribner's). Original work published 1937)

as successfully and innovatively working to build a more sustainability-as-flourishing future will experience higher levels of wellbeing as expressed by the key dimensions of PERMA.

The implications of this hypothesis are of course enormous. The phrase "Do good, do well" becomes more than a social responsibility mantra. Of all the things that can bring out the best in human beings, one of the most transformational is the mirroring effect that happens when we help bring out the best in Nature and in others. The reality of mirror flourishing, when it is experienced most authentically, might well be the human development business opportunity of our time, at least under certain conditions such as those found in whole-system change practice, for example, the Appreciative Inquiry Sustainable Design Factory, or what Illma Barros has termed "holistic-AI," where reflective practices powerfully punctuate intensive large-group planning and designing.[22]

In alignment with observations, what we are seeing emerge is an incomparable way to engage and motivate the entire workforce—where people come alive with purpose, meaning, hope, inspiration, and intrinsically motivated accomplishment. Mirror flourishing speaks to the unified and integral two-way flow between business and our world—this fundamental blurring of the boundaries of "in here" and "out there"—and the possibility that, when we help life "out there" to flourish, we cannot help but benefit ourselves as well. I'll never forget when, in a 500-person sustainable design summit at Fairmount Minerals, an employee team came up with a new business idea for a US$15 sand water filter.[23] Ultimately it would be deployed in 44 countries, saving lives all over the world. But it would also benefit the business; it was not so much charity as a win–win sustainable value proposition. And the cross-functional team was alive and inspired, filled with pride in the company. And like a resonant note in music, this kind of work "out there" reverberated all the way down the line into many aspects of the everyday work experience. Can you imagine people lining up droves to work for a sand-mining company? We're not talking Silicon Valley glitz and glamor. We're talking about sand-loader operators, engineers, finance

22   Laszlo, C., Brown, J.S., Ehrenfeld, J.R., Gorham, M., Pose, I.B., Robson, L., ... Werder, P. (2014). *Flourishing Enterprise: The New Spirit of Business*. Stanford, CA: Stanford University Press.

23   Cooperrider, D., & McQuaid, M. (2012). The positive arc of systemic strengths: how appreciative inquiry and sustainable designing can bring out the best in human systems. *Journal of Corporate Citizenship*, 46, 71-102.

specialists, and production supervisors. That's what's happening to this award-winning enterprise.

Outside of the sustainability literature, even without a name, there are now over 500 scientific studies on this doing good/doing well dynamic. Stephan Post summarized many of them in a book titled *Why Good Things Happen to Good People*, and argued that this doing good/doing well dynamic is "the most potent force on the planet."[24] If you engage in helping activities as a teen, for example, you will still be reaping health benefits 60 or 70 years later. But in all of this research there is not one study relating this to sustainability work, what happens to us when we restore Nature, or the human impact in those corporations leading the world in sustainable value. Obviously the possibilities are wide open and vast.

The reversal of so much active disengagement in the workplace, as well as depression and heartsickness in our culture at large, might well be easier to accomplish than we think. More than 125 million small and medium-size businesses operate across and around our blue planet. Imagine the positive mirror flourishing effect of millions of flourishing enterprise and shared purpose initiatives reverberating, scaling up, and amplifying. Imagine sustainability-as-flourishing and the predictable dynamic of mirror flourishing being actively harnessed as a massive human development leverage point. This is macro-obliquity in action—focusing our attention "out there" where the real or mutual benefit is realized "in here." Neuroscience research supports all of this and suggests that people act their way into believing rather than thinking their way into acting. "We *are* what we repeatedly do," said Aristotle, "Excellence, then, is not an act but a habit." And then the Dalai Lama then answers the next really important question: "Why do we actually need compassion?" And here he does not hesitate: "Because it benefits us."

This is the way of the Bodhisattva, says Sander Tideman, and the need for *both* an inner and outer focus. Compassion is not just a warm feeling of empathy, but a *practice* that involves the courage to act. When you reread this book, notice the key question it deals with—especially pages 132-134. The major achievement of this work, in my view, is that it is the best business explanation of the connection between inner and outer transformation I've ever seen.

Each one of us is a seed, a quiet promise, and through a combination of our engagement "out there" and our reflective practices "in here" we will

24 Post, S. (2007). *Why Good Things Happen to Good People*. New York, NY: Random House, Inc.

discover that business truly does have the opportunity to be one of the most generative, creative, and positive forces on the planet.[25] And by taking initiative to introduce shared purpose and triple value and societal leadership into your organization as this book charts out, you will quickly find that you are not alone. Many people today, like you, share the hope that the 21st century can become, without one moment of delay, an unprecedented era of innovation where businesses can excel, people can thrive, and nature can flourish. It's an incredible time to be alive. There is a sense of tremendous privilege and sense of vast potential.

At one point I asked the Dalai Lama to share his vision for our schools of management where millions and millions of young leaders will soon be making the billions of decisions that add up to tremendous impact. He turned playful. He replied: "I myself am not a good manager. If I were to manage something, it would end up in chaos. If you come to my room, everything is very untidy. [Laughter]" In essence, in his truly modest way, he was saying to the business leaders, "This is your field." But then he later went on to speak about the education of the heart, and the need for a radical reorientation away from our preoccupation of the self to a concern for the other, and emphasized that "I believe every individual has the potential make at least some contribution to the happiness and welfare of humanity." But the words that moved me the most were words that speak directly to our moment when he said: "The future is totally open, like empty space, like a feather blowing in the wind—anything is possible. The 21st century is just beginning."

David L. Cooperrider, PhD
University Distinguished Professor
Case Western Reserve University & Honorary Chair
David L. Cooperrider Center for Appreciative Inquiry
Stiller School of Business, Champlain College
July 14, 2016

---

25   Cooperrider, D.L., Zandee, D.P., Godwin, L.N., Avital, M., & Boland, B. (Eds.) (2013). *Organizational Generativity: The Appreciative Inquiry Summit and a Scholarship of Transformation.* Bingley, UK: Emerald Group Publishing Limited.

Participants

Compassion or Competition, Amsterdam, 1999

The Dalai Lama
His Holiness the 14th Dalai Lama, Tenzin Gyatso, is the spiritual leader of the Tibetan people. In 1959, a few years after the Chinese invasion of Tibet, he was forced to flee his homeland and seek asylum in neighboring India where he has lived ever since. In 1989 he was awarded the Nobel Prize for Peace in recognition of his nonviolent struggle to liberate Tibet, and in 2008 he received the U.S. Congressional Gold Medal, the highest civilian honor in the U.S.A. His simplicity and friendliness move everyone who meets him during his lectures and travels around the world. His message focuses on the importance of love, compassion, and forgiveness, which he refers to as "secular ethics" in the sense that they transcend religious boundaries and are relevant to all human endeavors.

Jermyn Brooks
Global Managing Partner of Price Waterhouse Coopers (1998–2002); Executive Director of Transparency International.

Wessel Ganzevoort
Professor of Economics, University of Amsterdam; former Chairman of KPMG Consulting.

Hazel Henderson
Author of *Building a Win–Win World* and five other highly acclaimed books.

Ruud Lubbers
Honorary Minister of State and former Prime Minister of the Netherlands (1982–94); former UN High Commissioner for Refugees; Commissioner of the Earth Charter.

Fred Matser
President of the Fred Foundation; director of SOFAM Beheer BV; humanitarian dedicated to a more functional society.

Stanislav Menchikov
Economist; author of *The Compassionate Economy*; Professor of Economics, Russian Federation.

Geoff Mulgan
Founder of Demos, author and visiting lecturer at the University of Westminster; former adviser to Prime Minister Tony Blair.

Hans Opschoor
Former Dean of the Institute of Social Studies, The Hague; author, economist.

Marcello Palazzi
Founder President of the Progressio Foundation; cofounder of B Corps Europe.

Erica Terpstra
Chairperson of NOC-NCF, the Dutch Olympic Committee; former minister of the Dutch Government.

Sander Tideman
International banker; cochair and cofounder of Spirit in Business.

Henk van Luijk
Former Professor of Business Ethics at Nyenrode University; founder of the European Business Ethics Network.

Eckart Wintzen
Founder President of Ex'tent BV; founder and former CEO of Origin.

Designing an Economy that Works for Everyone, Irvine, 2004

H.H. the Dalai Lama

W. Brian Arthur
Citibank Professor in Economics at the Santa Fe Institute.

Karen Wilhelm Buckley
Co-director and founder of the Wisdom Connection.

David L. Cooperrider
Professor in the Department of Organizational Behavior at the Weatherhead School of Management, Case Western Reserve University.

Betsy Dunham
Healthcare strategist.

Anders Ferguson
Partner at Veris Wealth Partners; cofounder of Spirit in Business.

Bill George
Professor of Management Practice at Harvard Business School; former CEO of Medtronic; author of *Authentic Leadership*.

John L. Graham
Professor of Marketing and International Business at the Merage School of Business, University of California.

Barbara J. Krumsiek
Chair, CEO, and President of Calvert Group, Ltd.

Peter Miscovich
Managing Director of Jones Lang LaSalle.

Terry Pearce
Founder and President of Leadership Communication.

Mark Thompson
Co-author of *Success Built to Last*; venture investor.

Leadership for a Sustainable World, The Hague, 2009

H.H. the Dalai Lama

Peter Blom
President and CEO of Triodos Bank.

Roger Dassen
Global Managing Partner of Deloitte.

Anders Ferguson
Chairman of Veris Wealth Management.

Li Hong Hui
Entrepreneur.

Ruud Lubbers
Honorary Minister of State and former Prime Minister of the Netherlands (1982–94); former UN High Commissioner for Refugees; Commissioner of the Earth Charter.

Sander Tideman
Cofounder and Managing Director of the Flow Foundation.

Roger van Boxtel
CEO of Menzis Health Insurance; former minister of the Dutch Government.

Irene van Lippe-Biesterfeld
Princess of the Netherlands; Chairperson of the Van Lippe-Biesterfeld Nature College

Herman Wijffels
Former CEO of Rabobank; former Governor of the World Bank; former Secretary General of the Dutch Social Economic Council (SER).

Education of the Heart, Rotterdam, 2014

H.H. the Dalai Lama

Ruud Lubbers
Honorary Minister of State and former Prime Minister of the Netherlands (1982–94); former UN High Commissioner for Refugees; Commissioner of the Earth Charter.

Gerdt Kernkamp
Founder and CEO of the School for Excellence.

Daniel Siegel
Professor of Psychiatry at the UCLA School of Medicine; Executive Director of the Mindsight Institute.

Mark Simons
Founder and CEO of Mountain Child Care.

Luuk Stevens
Professor; founder and Chairman of NIVOZ (Netherlands Institute for Educational Matters).

Geert ten Dam
Chairperson of VO Raad (Dutch Education Council).

Sander Tideman
Cofounder and Managing Director of the Flow Foundation.

Rob van Tulder
Professor of Business and Society Management at the Rotterdam School of Management, Erasmus University.

Mark Woerde
Founder and CEO of Lemz Advertisement Company.

Bibliography

Akerlof, G.A., & Schiller, R.J. (2008). *Animal Spirits: How Human Psychology Drives the Economy, and Why it Matters for Global Capitalism*. Princeton, NJ: Princeton University Press.

Akiner, S., Tideman, S.G., & Hay, J. (Eds.). (1998). *Sustainable Development in Central Asia*. London: Curzon Press.

Aknin, L., Norton, M., & Dunn, E. (2009). From wealth to well-being? Money matters, but less than people think. *The Journal of Positive Psychology*, 4(6), 523-527.

Ariely, D. (2008). *Predictably Irrational: The Hidden Forces That Shape Our Decisions*. New York, NY: HarperCollins.

———— (2013). *The Honest Truth About Dishonesty: How We Lie to Everyone—Especially Ourselves*. New York, NY: Harper Perennial.

Bakshi, R. (1998). *Bapu Kuti: Journeys in Rediscovery of Gandhi*. New Delhi, India: Penguin Books.

Barrett, R. (1998). *Liberating the Corporate Soul: Building a Visionary Organization*. London: Routledge.

———— (2009). *Building a Values-Driven Organization: A Whole-Systems Approach to Culture Transformation*. London: Butterworth-Heinemann.

Bass, B.M. (1998). *Transformational Leadership: Industrial, Military, and Educational Impact*. Mahwah, NJ: Lawrence Erlbaum Associates.

Beck, D.E., & Cowan, C.C. (1996). *Spiral Dynamics: Mastering Values, Leadership, and Change*. Cambridge, MA: Blackwell.

Bernanke, B. (2010). The economics of happiness. University of South Carolina commencement ceremony, Columbia, SC. May 8, 2010. Retrieved from http://www.federalreserve.gov/newsevents/speech/bernanke20120806a.htm.

Block, P. (1987). *The Empowered Manager*. San Francisco, CA: Jossey-Bass.

Boiral, O., Cayer, M., & Baron, C.M. (2009). The action logics of environmental leadership: a developmental perspective. *Journal of Business Ethics*, 85, 479-499.

Brown, B.C. (2011). *Conscious Leadership for Sustainability: How Leaders with a Late-Stage Action Logic Design and Engage in Sustainability Initiatives* (Unpublished PhD dissertation). Fielding Graduate University, Santa Barbara, CA.

Buber, M. (1958). *I and Thou* (New York: Scribner's). Original work published 1937)

Burns, J.M. (1978). *Leadership*. New York, NY: Harper & Row.

Carson, R. (2002). *Silent Spring*. Boston, MA: Houghton Mifflin. (Original work published 1962)

Christakis, N., & Fowler, J. (2009). *Connected: How Your Friends' Friends' Friends Affect Everything You Feel, Think, and Do*. New York, NY: Little, Brown, & Company.

Collins, J. (2001). *Good to Great: Why Some Companies Make the Leap. And Others Don't*. New York, NY: HarperBusiness.

Collins, J., & Porras, I. (1997). *Built to Last: Successful Habits of Visionary Companies*. New York, NY: HarperCollins.

Confino, J. (2013, October 2). Interview: Unilever's Paul Polman on diversity, purpose and profits. *The Guardian*. Retrieved from http://www.theguardian.com/sustainable-business/unilver-ceo-paul-polman-purpose-profits.

Cook-Greuter, S.R. (1999). Postautonomous ego development: a study of its nature and measurement. *Dissertation Abstracts International*, 60 06B (UMI No. 993312).

———— (2004). Making the case for a developmental perspective. *Industrial and Commercial Training*, 36(7), 277-281.

Cooperrider, D. (2012). The concentration effect of strengths. *Organizational Dynamics*, 42(2), 21-32.

Cooperrider, D., & Fry, R. (2012). Mirror flourishing and the positive psychology of sustainability. *Journal of Corporate Citizenship*, 46, 3-12.

Cooperrider, D., & Godwin, L. (2011). Positive organization development. In K. Cameron & G. Spreitzer (Eds.), *The Oxford Handbook of Positive Organizational Scholarship* (pp. 737-750). Oxford, UK: Oxford University Press.

Cooperrider, D., & McQuaid, M. (2012). The positive arc of systemic strengths: how appreciative inquiry and sustainable designing can bring out the best in human systems. *Journal of Corporate Citizenship*, 46, 71-102.

Cooperrider, D.L., Zandee, D.P., Godwin, L.N., Avital, M., & Boland, B. (Eds.) (2013). *Organizational Generativity: The Appreciative Inquiry Summit and a Scholarship of Transformation*. Bingley, UK: Emerald Group Publishing Limited.

Csikszentmihalyi, M. (1990.) *Flow: The Psychology of Optimal Experience*. New York, NY: Harper & Row.

Dalai Lama (2000). *Ethics for the New Millennium*. New York, NY: Riverhead Books.

———— (2002). *The Meaning of Life from a Buddhist Perspective*. Boston, MA: Wisdom Publications.

———— (2005). *The Universe in a Single Atom*. New York, NY: Random House.

Dalai Lama, & Norman, A. (2011). *Beyond Religion: Ethics for a Whole World*. New York, NY: Houghton Mifflin Harcourt Publishing.

Dalai Lama, & Ouaki, F. (1999). *Imagine All the People: A Conversation with the Dalai Lama on Money, Politics, and Life as it Could Be*. Boston, MA: Wisdom Publications.

Dalai Lama, & van den Muyzenberg, L. (2009). *The Leader's Way: The Art of Making the Right Decisions in Our Careers, Our Companies, and the World at Large*. New York, NY: Crown Business.

Davidson, R.J., & Begley, S. (2012). *The Emotional Life of Your Brain: How its Unique Patterns Affect the Way You Think, Feel, and Live—and How You Can Change Them*. New York, NY: Hudson Street Press.

Deci, E.L., & Ryan, R. (2000). The self-determination theory, the facilitation of intrinsic motivation, social development, and well-being. *American Psychologist*, 55(1), 68-78.

Deloitte Sports Business Group (2016). *Top of the Table: Football Money League*. Retrieved from http://www2.deloitte.com/content/dam/Deloitte/uk/Documents/sports-business-group/uk-deloitte-sport-football-money-league-2016.pdf.

Dixon, F. (forthcoming). *Global System Change: Achieving Sustainability and Real Prosperity*. In press.

Doppelt, B. (2009). *Leading Change Toward Sustainability: A Change-Management Guide for Business, Government and Civil Society* (2nd ed.). Sheffield, UK: Greenleaf Publishing.

———— (2012). *The Power of Sustainable Thinking: How to Create a Positive Future for the Climate, the Planet, Your Organization and Your Life.* London: Earthscan.

Doyle, E. (1996). *St. Francis and the Song of Brotherhood and Sisterhood.* New York, NY: Franciscan Institute.

DSM (2013, June 8). DSM to contribute to new 2020 global nutrition target. Retrieved from http://www.dsm.com/markets/paint/en_US/news-events/2013/06/14-13-dsm-to-contribute-to-new-2020-global-nutrition-target.html.

Dweck, C.S. (2006). *Mindset: How You Can Fulfill Your Potential.* London: Robinson.

Easterlin, R. (2008). Income and happiness: towards a unified theory. *The Economic Journal,* 11(473), 465-484.

Eccles, R.G., Ioannou, I., & Serafeim, G. (2014). The impact of corporate sustainability on organizational processes and performance. *Management Science,* 60(11), 2835–2857. Retrieved from https://dash.harvard.edu/handle/1/15788003.

The Economist (2016, February 20). The world economy: out of ammo? *The Economist,* 36.

Edelman Berland (2014). 2014 Edelman Trust Barometer. Retrieved from http://www.edelman.com/insights/intellectual-property/2014-edelman-trust-barometer.

Edgeworth, F.Y. (1967). *Mathematical Psychics: An Essay on the Application of Mathematics to the Moral Science.* New York, NY: Augustus M. Kelley Publishers. (Original work published 1881).

Eigendorf, J. (2009, July 16). Dalai Lama—"I am a supporter of globalization". *Die Welt.* Retrieved from http://www.welt.de/english-news/article4133061/Dalai-Lama-I-am-a-supporter-of-globalization.html.

Elias, N. (2000). *The Civilizing Process: Sociogenetic and Psychogenetic Investigations.* Oxford, UK: Basil Blackwell.

Epstein, M.J. (2014). *Making Sustainability Work: Best Practices in Managing and Measuring Corporate Social, Environmental, and Economic Impacts.* Sheffield, UK: Greenleaf Publishing.

Financial Times (2016). Equities. Retrieved from http://markets.ft.com/research/Markets/Tearsheets/Financials.

Fisher, D., Rooke, D., & Torbert, W. (2003). *Personal and Organizational Transformations Through Action Inquiry.* London: The Cromwell Press.

Fortune (2009). The 10 largest U.S. bankruptcies. *Fortune.* Retrieved from http://archive.fortune.com/galleries/2009/fortune/0905/gallery.largest_bankruptcies.fortune/.

Forum for the Future (2011). Interview with Paul Polman. Retrieved from https://www.forumforthefuture.org/blog/6-ways-unilever-has-achieved-success-through-sustainability-and-how-your-business-can-too.

Fox, E.A. (2013). *Winning From Within: Breakthrough Method for Leading, Living, and Lasting Change.* New York, NY: HarperCollins.

Fredrickson, B.L. (2003). The value of positive emotions. *American Scientist,* 91, 330-335.

Freeman, R.E. (1984). *Strategic Management: A Stakeholder Approach.* Boston, MA: Pitman.

Friedman, M. (1970, September 13). The social responsibility of business is to increase its profits. *New York Times,* p. SM17.

George, B. (2003). *Authentic Leadership: Rediscovering the Secrets of Creating Lasting Value.* San Francisco, CA: Jossey-Bass.

Gilding, P. (2011). *The Great Disruption: How the Climate Crisis Will Transform the Global Economy.* London: Bloomsbury.

Gintis, H. (2000). Beyond homo economicus: evidence from experimental economics. *Ecological Economics,* 35, 311-322.

Gitsham, M. (2009). *Developing the Global Leader of Tomorrow*. Berkhamsted, UK: Ashridge Business School.

Goleman, D. (1995). *Emotional Intelligence: Why It Can Matter More Than IQ for Character, Health and Lifelong Achievement*. New York, NY: Bantam Books.

—— (1997). *Healing Emotions: Conversations with the Dalai Lama on Mindfulness, Emotions, and Health*. Boston, MA: Shambhala.

—— (2003). *Destructive Emotions: How Can We Overcome Them? A Scientific Dialogue with the Dalai Lama*. New York, NY: Bantam Dell.

Goleman, D., McKee, A., & Boyatzis, R. (2002). *Primal Leadership: Leading with Emotional Intelligence*. Boston, MA: Harvard Business Press.

Gowdy, J. (2009). *Economic Theory Old and New: A Student's Guide*. Palo Alto, CA: Stanford University Press.

Hamel, G., & Breen, B. (2007). *The Future of Management*. Cambridge, MA: Harvard Business School Publishing.

Hames, R.D. (2007). *The Five Literacies of Global Leadership: What Authentic Leaders Know and You Need to Find Out*. Hoboken, NJ: Jossey-Bass.

Hardman, G. (2009). *Regenerative Leadership: An Integral Theory for Transforming People and Organizations for Sustainability in Business, Education, and Community* (Unpublished PhD dissertation). Florida Atlantic University, Boca Raton, FL.

Henrich, J., Boyd, R., Bowles, S., Camerer C., Fehr E., & Gintis H. (2004). *Foundations of Human Sociality: Economic Experiments and Ethnographic Evidence from Fifteen Small-Scale Societies*. Oxford, UK: Oxford University Press.

Hind, P., Wilson, A., & Lenssen, G. (2009). Developing leaders for sustainable business. *Corporate Governance*, 9(1), 7-20.

Hollensbe, E., Wookey, C., Hickey, L., & George, G. (2014). Organizations with purpose. *Academy of Management Journal*, 57(5), 1227-1234.

Horwitz, F.M., & Grayson, D. (2010). Putting sustainability into practice in a business school. *Global Focus*, 4(2), 26-29.

Hougaard, R., Carter, J., & Coutts, G. (2015). *One Second Ahead: Enhance Your Performance at Work with Mindfulness*. London: Palgrave Macmillan.

Institute of Directors of Southern Africa (2009). King report on governance for South Africa 2009. Johannesburg: Institute of Directors of Southern Africa. Retrieved from https://jutalaw.co.za/uploads/King_III_Report/#p=1.

Johansen, B. (2012). *Leaders Make the Future: Ten Leadership Skills for an Uncertain World*. San Francisco, CA: Berrett-Koehler Publishers.

Kabat-Zinn, J. (1990). *Full Catastrophe Living: Using the Wisdom of Your Body and Mind to Face Stress, Pain, and Illness*. New York, NY: Bantam Books.

Kahneman, D. (1979). Prospect theory: an analysis of decision under risk. *Econometrica*, 47, 263-291.

Kahneman, D., Diener, E., & Schwarz, N. (Eds.) (1999). *Well-being: The Foundations of Hedonic Psychology*. New York, NY: Russell Sage Foundation.

Kakabadse, N.K., Kakabadse, A.P., & Lee-Davies, L. (2009). CSR leadership road-map. *Corporate Governance*, 9(1), 50-57.

Kegan, R. (1982). *The Evolving Self: Problem and Process in Human Development*. Cambridge, MA: Harvard University Press.

—— (1994). *In Over Our Heads: The Mental Demands of Modern Life*. Cambridge, MA: Harvard University Press.

Keynes, J.M. (1930). Economic possibilities for our grandchildren. In J.M. Keynes (Ed.), *The Collected Writings of John Maynard Keynes, Volume IX* (pp. 329-331). London: Macmillan.

Ki-moon, B. (2012, April 2). Secretary-General, in message to meeting on "Happiness and Well-being" calls for "Rio+20" outcome that measures more than Gross National Income. UN Press Release. Retrieved from http://www.un.org/press/en/2012/sgsm14204.doc.htm.

Kiron, D. (2013). The benefits of sustainability-driven innovation. *MIT Sloan Management Review*, 54(2), 69-73.

Kolk, A., & Perego, P. (2014). Sustainable bonuses: sign of corporate responsibility or window dressing? *Journal of Business Ethics*, 119(1), 1-15.

Kotter, J.P. (1996). *Leading Change*. Boston, MA: Harvard Business School Press.

Kouzes, J.M., & Posner, B.A. (1995). *The Leadership Challenge*. San Francisco, CA: Jossey-Bass.

Kuhn, T.S. (1962). *The Structure of Scientific Revolutions*. Chicago, IL: University of Chicago Press.

Lane, J. (2015, September 13). DSM, KLM rank high in annual Dow Jones Sustainability Index. *Biofuels Digest*. Retrieved from http://www.biofuelsdigest.com/bdigest/2015/09/13/dsm-klm-rank-high-in-annual-dow-jones-sustainability-index/.

Laszlo, C., Brown, J.S., Ehrenfeld, J.R., Gorham, M., Pose, I.B., Robson, L., ... Werder, P. (2014). *Flourishing Enterprise: The New Spirit of Business*. Stanford, CA: Stanford University Press.

Levy, D.M., & Peart, S.J. (2001). The secret history of the dismal science. Part I. Economics, religion and race in the 19th century. Library of Economics and Liberty. Retrieved from http://www.econlib.org/library/Columns/LevyPeartdismal.html.

Limb, C.J., & Braun, A.R. (2008). Neural substrates of spontaneous musical performance: an fMRI study of jazz improvisation. *PLoS ONE*, 3(2): e1679. doi:10.1371/journal.pone.0001679.

Loehr, J., & Schwartz, T. (2004). *The Power of Full Engagement*. New York, NY: Simon & Schuster.

Loevinger, J. (1976). *Ego Development: Conceptions and Theories*. San Francisco, CA: Jossey-Bass.

Loizzo, J. (2006). Renewing the Nalanda legacy: science, religion and objectivity in Buddhism and the West. *Religion East & West*, 6, 101-121.

——— (2012). *Sustainable Happiness: The Mind Science of Well-being, Inspiration and Compassion*. New York, NY: Routledge.

Lueneburger, C., & Goleman, D. (2010). The change leadership sustainability demands. *MIT Sloan Management Review*, 51(4), 49-55.

Lutz, A., Dunne, J., & Davidson, R. (2007). Meditation and the neuroscience of consciousness. In P. Zelazo, M. Moscovitch, & E. Thompson (Eds.), *Cambridge Handbook of Consciousness* (pp. 499-554). Cambridge, UK: Cambridge University Press.

Lutz, A., Slagter, H.A., Rawlings, N.B., Francis, A.D., Greischar, L.L., & Davidson, R.J. (2009). Mental training enhances attentional stability: neural and behavioral evidence. *The Journal of Neuroscience*, 29(42), 13418-13427.

Lyubomirsky, S., King, L., & Diener, E. (2005). The benefits of frequent positive affect: does happiness lead to success? *Psychological Bulletin of American Psychological Association*, 131(6), 803-855.

Mace, M. (2016, May 5). Elon Musk: "We need a revolt against the fossil fuel industry." *The Guardian*. Retrieved from https://www.theguardian.com/environment/2016/may/05/elon-musk-we-need-a-revolt-against-the-fossil-fuel-industry.

Mackay, J., & Sisodia, R. (2013). *Conscious Capitalism: Liberating the Heroic Spirit of Business*. Boston, MA: Harvard Business Review Press.

Marshall, J., Coleman, G., & Reason, P. (2011). *Leadership for Sustainability: An Action Research Approach*. Sheffield, UK: Greenleaf Publishing.

Martínez, R. (2010, July 11). World Cup final: Johan Cruyff sowed seeds for revolution in Spain's fortunes. *The Daily Telegraph*.

Maslow, A.H. (1943). A theory of human motivation. *Psychological Review*, 50(4), 370-396.

McClelland, D.C. (1978). Managing motivation to expand human freedom. *American Psychologist*, 33(3), 201-210.

Meadows, D.H., Meadows, D.L., Randers, J., & Behrens, W.W. (1972). *The Limits to Growth: A Report for the Club of Rome*. New York, NY: Universe Books.

Metcalf, L., & Benn, S. (2013). Leaders for sustainability: an evolution of leadership ability. *Journal of Business Ethics*, 112(3), 369-384.

Mirvis, P. (2011). Unilever's drive for sustainability and CSR—changing the game. In S.A. Mohrman, & A.B. Shani (Eds.), *Organizing for Sustainability Volume 1* (pp. 41-72). Bingley: Emerald Group Publishing Limited.

Mullin, G.H. (2001). *The Fourteen Dalai Lamas: A Sacred Legacy of Reincarnation*. Santa Fe, NM: Clear Light Publishers.

Nidumolu, R., Kramer, K., & Zeitz, J. (2012). Connecting heart to head: a framework for sustainable growth. *Stanford Social Innovation Review*, Winter. Retrieved from http://www.ssireview.org/articles/entry/connecting_heart_to_head.

NORAD (1999). *The Logical Framework Approach (LFA): Handbook for Objectives-Oriented Planning* (4th ed.). Oslo, Norway: NORAD.

Peterson, C., & Seligman, M. (2004). *Character Strengths and Virtues: A Handbook and Classification*. Washington, DC: APA Press and Oxford University Press.

Porter, M., & Kramer, M. (2006). Strategy & society: the link between competitive advantage and corporate social strategy. *Harvard Business Review*, 83(4), 66-67.

——— (2011). The big idea: creating shared value. *Harvard Business Review*, 1 (January–February), 64-77.

Post, S. (2007). *Why Good Things Happen to Good People*. New York, NY: Random House, Inc.

Proust, M. (1923). *Remembrance of Things Past*. Vol. 5: *The Captive* (C.K. Moncrief, Trans.). Retrieved from http://gutenberg.net.au/ebooks03/0300501.txt.

Quinn, L., & Dalton, M. (2009). Leading for sustainability: implementing the tasks of leadership. *Corporate Governance*, 9(1), 21-38.

Ray C. Anderson Foundation (n.d.). *Climbing Mount Sustainability*. Retrieved from http://www.raycandersonfoundation.org/assets/pdfs/rayslife/EssayClimbingMountSustainability.pdf.

Schein, S. (2015). *A New Psychology for Sustainability Leadership: The Hidden Power of Ecological Worldviews*. Sheffield, UK: Greenleaf Publishing.

Seligman, M.E.P. (2002). *Authentic Happiness: Using the New Positive Psychology to Realize Your Potential for Lasting Fulfillment*. New York, NY: Free Press.

—— (2010). *Flourish: A Visionary New Understanding of Well-Being*. New York, NY: Free Press.

Seligman, M., & Csikszentmihalyi, M. (2000). Positive psychology: an introduction. *American Psychologist*, 55, 5-14.

Senge, P.M. (1990). *The Fifth Discipline: The Art and Practice of the Learning Organization*. New York, NY: Doubleday Currency.

——— (2008). *The Necessary Revolution: How Individuals and Organizations Are Working Together to Create a Sustainable World*. New York, NY: Broadway Books.

Senge, P., Sharmer, O., Jaworski, J., & Flowers, B.S. (2008). *Presence: Human Purpose and the Field of the Future*. Boston, MA: Crown Business.

Shantideva (1997). *A Guide to the Bodhisattva Way of Life* (V.A. Wallace & B.A. Wallace, Trans.). Ithaca, NY: Snow Lion.

Sharma, S. (2000). Managerial interpretations and organizational context as predictors of corporate choice of environmental strategy. *Academy of Management Journal*, 43(4), 681-697.

Sharmer, C.O. (2008). *Theory U: Leading from the Future as it Emerges*. San Francisco, CA: Berrett-Koehler Publishers.

Shrivastava, P. (1994). Ecocentric leadership in the 21st century. *The Leadership Quarterly*, 5(3), 223-226.

Siegel, D. (2009). *Mindsight: The New Science of Personal Transformation*. New York, NY: Random House.

————— (2013). *The Developing Mind: How Relationships and the Brain Interact to Shape Who We Are*. New York, NY: Random House.

Sijbesma, F. (2013). We need to redesign our economy. *Huffington Post Business*. Retrieved from http://www.huffingtonpost.com/feike-sijbesma/we-need-to-redesign-our-e_b_2597564.html.

Singer, T. (2008). Understanding others: brain mechanisms of theory of mind and empathy. In P.W. Glimcher, C.F. Camerer, E. Fehr, & R.A. Poldrack (Eds.), *Neuroeconomics: Decision Making and the Brain* (pp. 233-250). Amsterdam, Netherlands: Elsevier.

Singer, T., & Ricard, M. (2015). *Caring Economics: Conversations on Altruism and Compassion Between Scientists, Economists, and the Dalai Lama*. New York, NY: Picador.

Sisodia, R.S., Wolfe, D.B., & Sheth, J.N. (2007). *Firms of Endearment: How World-Class Companies Profit from Passion and Purpose*. New York, NY: Prentice Hall.

Smith, A. (2005). *The Theory of Moral Sentiments* (S.M. Soares, Ed.). São Paulo, Brazil: Metalibri. (Original work published 1790).

————— (2008). *The Wealth of Nations*. New York, NY: Oxford University Press. (Original work published 1776).

Sogyal, R. (2002). *The Tibetan Book of Living and Dying*. New York, NY: HarperCollins.

Stebnicki, M.A. (2007). *Empathy Fatigue: Healing the Mind, Body, and Spirit of Professional Counselors*. New York, NY: Springer.

Suresh, S., & Cooperrider, D.L. (1998). *Organizational Wisdom and Executive Courage*. Lanham, MD: Lexington Press.

Tan, C.-M. (2012). *Search Inside Yourself: The Unexpected Path to Achieving Success, Happiness (and World Peace)*. New York, NY: HarperOne.

Thaler, R.H., & Sunstein, C.R. (2008). *Nudge: Improving Decisions About Health, Wealth, and Happiness*. New Haven, CT: Yale University Press.

Thurman, R. (1997). *Inner Revolution: Life, Liberty and the Pursuit of Real Happiness*. New York, NY: Riverhead Books.

Tideman, S. (2009). *Mind Over Matter: Van Zeepbelkapitalisme naar Economie met een Hart*. Amsterdam: Business Contact Press.

————— (2016). Gross national happiness: lessons for sustainability leadership. *South Asia Journal for Global Business Research*, 5(2), 190-213.

Tideman, S.G., Arts, M.C.L., & Zandee, P.D. (2013). Sustainable leadership: towards a workable definition. *Journal of Corporate Citizenship*, 49, 17-33.

Torbert, W.R., Cook-Greuter, S.R., Fisher, D., Foldy, E., Gauthier, A., Keeley, J., ... Tran, M. (2004). *Action Inquiry: The Secret of Timely and Transforming Leadership*. San Francisco, CA: Berrett-Koehler Publishers.

Trungpa, C. (1984). *Shambhala: The Sacred Path of the Warrior*. Boulder, CO: Shambhala Publications.

Unilever (2012). *Unilever Sustainable Living Plan: Progress Report 2012*. Retrieved from https://www.unilever.com/Images/uslp-progress-report-2012-fi_tcm13-387367_tcm244-409862_en.pdf.

————— (2016). Sustainable living. Retrieved from https://www.unilever.com/sustainable-living.

United Nations (2012, July 12). Resolution 66/281 on Happiness. New York, NY: UN General Assembly.

————— (no date). Sustainable Development Knowledge Platform. Retrieved from https://sustainabledevelopment.un.org.

Ura, K., & Galay, K. (Eds.) (2004). *Gross National Happiness and Development*. Thimphu, Bhutan: Center for Bhutan Studies. Retrieved from http://www.bhutanstudies.org.bt/category/conference-proceedings/.

van Tulder, R. (2012). *Skill Sheets: An Integrated Approach to Research, Study and Management*. Amsterdam, Netherlands: Pearson.

van Tulder, R., van Tilburg, R., Francken, M., & de Rosa, A. (2014). *Managing the Transitions to Sustainable Enterprise: Lessons from Frontrunner Companies*. London: Earthscan/Routledge.

Varela, F., Thompson, E.T., & Rosch, E. (1992). *The Embodied Mind: Cognitive Science and Human Experience*. Boston, MA: MIT Press.

Visser, W. (2010). The age of responsibility: CSR 2.0 and the new DNA of business. *Journal of Business Systems, Governance and Ethics*, 5(3), 7-22.

Wallace, B.A. (1993). *Buddhism from the Ground Up*. Boston, MA: Wisdom Publications.

———— (2006). *Contemplative Science: Where Buddhism and Neuroscience Converge*. New York, NY: Columbia University Press.

WBCSD (2011). *Collaboration, Innovation and Transformation: A Value Chain Approach*. Geneva: World Business Council for Sustainable Development.

———— (2012). *Measuring Impact Framework: A Guide for Business*. Geneva: World Business Council for Sustainable Development.

Weick, K. (1979). *The Social Psychology of Organizing*. New York, NY: Addison-Wesley.

White, A.G. (2007). A global projection of subjective well-being: a challenge to positive psychology? *Psychtalk*, 56, 17-20.

Williams, F. (2015). *Green Giants: How Smart Companies Turn Sustainability into Billion-Dollar Businesses*. New York, NY: AMACOM.

World Bank (2014). *Food Price Watch*, 16. Retrieved from http://www-wds.worldbank.org/external/default/WDSContentServer/WDSP/IB/2014/05/30/000470435_20140530113813/Rendered/PDF/883900NEWS0Box000FPW0Feb020140final.pdf.

World Commission on Environment and Development (1987). *Our Common Future*. New York, NY: United Nations.

Zak, P.J. (2008). *Moral Markets: The Critical Role of Values in the Economy*. Princeton, NJ: Princeton University Press.

About the author

Sander Tideman became an expert in leadership development and sustainable business after a successful career as banker and business consultant. He has worked with organizations across the world and is Managing Director of Mind & Life Europe, a faculty member at the Department of Business-Society Management at Rotterdam School of Management, Erasmus University, and a cofounding partner of the Flow Foundation and Flow Impact Fund. After a meeting  with the Dalai Lama as a young man, he developed a lifelong friendship that would prove to be very significant in informing his thinking. Over the course of many years, he has engaged in a number of dialogues with the Dalai Lama and other leaders in business, society, and education.

FlowImpactFund

The Flow Impact Fund is dedicated to bringing the ideas described in this book into the world. It supports research and education projects and generates awareness for the development of societal business leadership to create triple value with organizations and society. Specifically, it focuses on the creation of educational modules, know-how, practical tools, and metrics for leaders within the framework of "Winning with Society."

The Flow Impact Fund also supports social impact projects for young personal leadership development in countries such as Bhutan, Tibet, and Nepal, in conjunction with offering educational triple value leadership journeys to these countries for business leaders.

For more information, please see http://www.flowimpactfund.nl.

Leadership journey in the Himalayas.